AMERICAN
STABLES

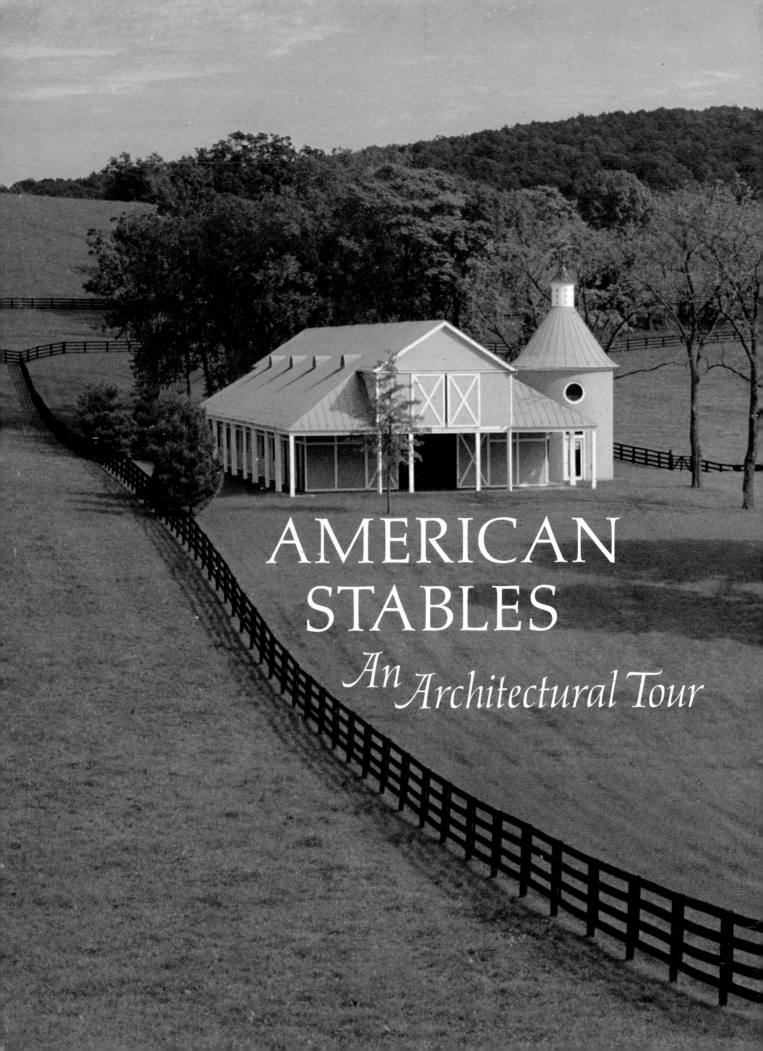

AMERICAN STABLES

An Architectural Tour

Julius Trousdale Sadler, Jr.
Jacquelin D. J. Sadler
Foreword by Alexander Mackay-Smith

NEW YORK GRAPHIC SOCIETY · BOSTON

Absent friends!

Color plate 1 *(half title). Stables, Saratoga Racetrack, Saratoga Springs, New York.*

Color plate 2 *(title page). Broodmare barn, Catoctin Stud, Waterford, Virginia.*

Color plate 3 *(copyright page). Pegasus, by Carl Milles. Meadowbrook Hall, Michigan State University, Rochester.*

Copyright © 1981 by Julius Trousdale Sadler, Jr., and Jacquelin D. J. Sadler

First Edition

Designed by Betsy Beach

New York Graphic Society books are published by Little, Brown and Company.
Published simultaneously in Canada by Little, Brown and Company (Canada) Limited.

Printed in the United States of America.

Library of Congress Cataloging in Publication data will be found on the last page.

Contents

Books by Julius Trousdale Sadler, Jr.

With Desmond Guinness
Mr. Jefferson, Architect
Palladio: A Western Progress

With Calder Loth
The Only Proper Style: Gothic Architecture in America

Yearlings, Catoctin Stud, Waterford, Virginia.

Foreword

New York Graphic Society is to be congratulated on the publication of this book, which fills a long-standing vacancy in the history of American architecture. We have had a plethora of volumes picturing the great mansions of the North built by mercantile and manufacturing fortunes, and of the even larger mansions of the South built by agricultural fortunes, volumes devoted almost entirely to the principal dwelling houses, with little or no mention of the dependency buildings that made them functional. During the summers in the South, before the days of screens and refrigeration, an outside summer kitchen building was essential. Where business transactions with outsiders were frequent, and where the volume of business required extensive bookkeeping, a separate office building was a matter of great convenience. On large and remote farms and plantations the employment of a tutor for the children of the "Great House" and for the children of the neighbors necessitated the building of a schoolhouse with one or two classrooms on the ground floor and an apartment for the tutor above. Wash houses, spring houses, and dairy houses were also standard features, as well as housing for servants, black and white. These various buildings, often charming in themselves, gave added distinction to the mansion house and were an important part of its surrounding landscape.

Before the coming of the automobile some three quarters of a centry ago, transportation and farm power were centered in the horse. Without riding and driving horses the occupants of the mansion were housebound. It follows that stables and carriage houses were the most universal of all the accompanying outbuildings. They were of equal importance in the mercantile and industrial cities of the North. Private stables housed the broughams and barouches of the leading citizens, while public livery stables served as the headquarters for hansom cabs and other vehicles for hire. All types of goods and merchandise were transported by drays and delivery wagons, while streetcars and buses were also horse-drawn.

For horsemen and for students of architecture Mr. and Mrs. Sadler have presented a particularly handsome and interesting collection of photographs—of accommodations for horses and carriages, from the seventeenth-century stableyard of the Black Horse Inn in Philadelphia to the mid-1970s Thoroughbred stables of River Edge Farm in California. The accompanying text, lively and well written, provides full backgrounds, equine and architectural. No period or area is slighted—the horse buildings of the last three centuries designed for domestic use and for sport, competitive and otherwise, are amply considered. Later editions could easily include an even broader spectrum without sacrificing the interest and quality of this pioneer publication.

ALEXANDER MACKAY-SMITH
"Woodley"
Berryville, Virginia

An Edwardian carriage house, stable, and indoor school in Dutchess County, New York. Rescue came too late to save the major portion of this once graceful building.

viii

Preface

American stabling runs the same gamut that American houses do, and varies as widely with income, purpose, climate, period, and local custom, from the run-in shed at one end of the spectrum to the equine equivalent of a château at the other. Many stable buildings, whether intended for a score of blood horses or a solitary backyard pony, have a considerable impact in terms of site, aesthetic, and use. As a design form, the genre and its place in our architectural heritage have received too little attention. Soon it may be too late to look. Very few of even the grandest stables and carriage houses of the past survive in their original form, in spite of the present remarkable increase in the equine population, because their world has altered around them. At the same time, a whole industry has grown up to supply the demand for functional, seemly stables and covered riding arenas at a reasonable cost. Many of these new buildings are attractive as well as workmanlike, but there is an inevitable uniformity inherent in mass production and design.

Exterior simplification has of course been accompanied by a sad falling off in interior decor. Vaulted ceilings, intricately laid brick aisles, ornamental woodwork, and decorative fittings have gone the way of the dinosaur. They are redundant to our modern pragmatism. One regrets the lost hardware: graceful cast-iron harness hooks, pillared stall posts capped with horsehead, pineapple, or simple ball finials, cunningly contrived brass stall latches (proof against the wiliest pony), mahogany blanket rails on scrolled and hinged supports, patterned grilles for stall doors and partitions. We and our horses do very well without them, but what pleasing flourishes they were, and what an air of solid worth and Victorian domesticity they still lend where they survive today.

In the course of research for his previous books, Mr. Sadler compiled an extensive file of intriguing stables and carriage houses which remained unused. (No matter how many pictures there may be in a work on architecture—where the image is, after all, most of the message—they represent the result of a rigorous and somewhat painful selection process. Even the most indulgent publisher is bound by considerations of space and economy.) It may be fairly said that our tour began among this legion of the disallowed. Starting in Europe and moving

1

with the sun, we have journeyed across many thousands of miles and half a millennium, following the history of American stabling in its various periods, climates, and circumstances.

The rewards have been great. In the endless reaches of prairie and desert, along moss-hung byways in the tidewater, steep county roads in New England, and multilane freeways on the West Coast, we have shared the freemasonry of horsepeople. Owners and managers have been patiently generous in showing us their establishments and in recommending others. With their help, we have tried to trace the evolution of the American stable—as distinguished from the barn, the byre, and the fold—using one or more of the following criteria: The building is, or was, of particular structural interest, designed by a noted architect, especially innovative or effective in plan, successfully adapted to a new use, the home of eminent horses, or simply especially pleasing to the eye.

We have left out a number of favorites because of their familiarity, and other interesting examples at the request of the owners; we have no doubt failed to discover any number of splendid stables. Perhaps these inevitable omissions will encourage readers to make their own journeys of discovery. Sta-

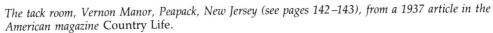

The tack room, Vernon Manor, Peapack, New Jersey (see pages 142–143), from a 1937 article in the American magazine Country Life.

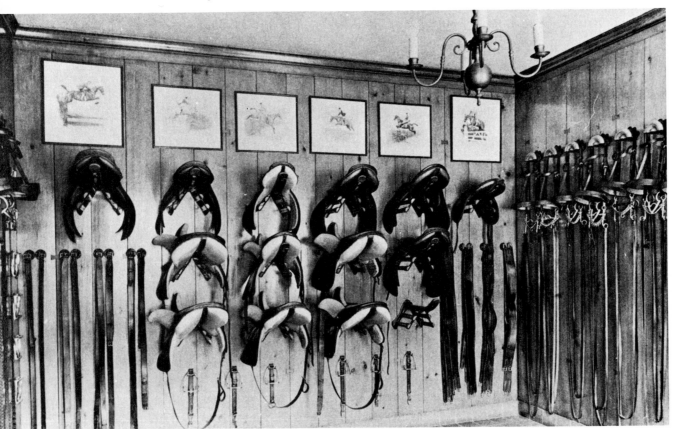

2

A stable in the Genesee hunt country, New York, displaying mid-nineteenth-century chic Gothic detail.

bling, like other architecture, reveals a great deal about its creators and their times. Moreover, it continues to reflect the ancient, enduring, and affectionate tie between horse and human.

Having tried to lay the blame for any oversight to the account of editorial ruthlessness and parsimony, the authors would like to express their gratitude for the encouragement, patience, and support faithfully extended to them by those involved in the production of this book.

3

The earliest Chinese character for "wheel" in the Oracle Bone script of the Shang Dynasty, ca. 1766–1121 B.C. The ideograph represents a pair-horse chariot seen from above.

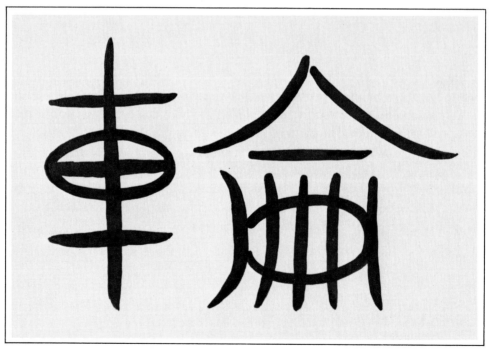

I Creature Comforts

You ought to build a lodge . . . to defend them from the inclemencies of the weather; for there is no creature to which cold is more injurious than to horses, neither will they endure much heat.

—WILLIAM CAVENDISH, DUKE OF NEWCASTLE

The beginnings of the association between horse and human are prehistoric. We are accustomed to think of the proto-Achaean nomad, archetype of the mythic centaur, as the first horseman. Perhaps he was the first mounted horseman, but the earliest tangible evidence of equine domestication is Chinese. Remains of horse chariots have been discovered in northern China dating from 3500 B.C.—more than a thousand years before the fierce little riders of the steppe swept down on the Greek peninsula. These shielded vehicles suggest that horses found military employment from the outset, although the first warhorses, it seems, were not troopers but draft animals, serving in harness.

Early Sumerian, Indian, and Egyptian reliefs show horses sharing peaceful pursuits with their masters. But cavalry and horse-drawn engines of war soon became an important component of every major army, and so remained for thousands of years. Only after the First World War, with the refinement of that last logical rediscovery of the chariot, the armored tank, was the horse at last honorably retired from active service on the battlefield; only in this century did the dubious victory of the internal combustion engine make mortal horsepower redundant in our daily lives. The true working horse has always kept a modest place, even in the automated society, notably where religious scruples or severe terrain inhibit the use of motorized equipment, and is also found in mounted police troops, or drawing some anachronistic commercial vehicle. The draft teams that vie with each other in shifting monstrous loads at country fairs often keep fit and earn their keep by working on the farm; indeed, there is a revival of interest in the horse as an agriculture aide. But in general we no longer make war with horses, nor cultivate the land with them, nor travel behind them or on their backs, nor forward mail and freight with their aid. Yet the horse is once more a factor in the national economy. As playthings and leisure companions, horses have become almost as important as when they provided the greater part of our motive power.

The age of the machine has given us more and more time to spend as we choose. The number of persons who devote large portions of that time to horses tells a lot about how hard many people are willing to work in the name of pleasure. Watching horses perform, or simply betting on them, also has innate appeal. Thoroughbreds, Standardbreds, Quarter Horses, Arabians, other light horses, and ponies race on the flat, over obstacles, and in harness in more than half our states, forming a great American industry.

Horse shows, once of a merely parochial interest, have multiplied. (More than eight thousand annual shows have recognized divisions for the American Saddle Horse alone.) The eighteenth-century sport of foxhunting flourishes in the face of rocketing costs and shrinking open land. Dressage, disdained in this country for many years as "circus work," is now recognized as a rewarding and elegant discipline, while the erstwhile military Combined Training Event (Dressage, Test in the Open, and Show Jumping) is now a growing and widely supported civilian sport. Competitive Trail and Endurance Rides have spread from the Sierra to the ancient mountain systems of the East, wherever suitable terrain can be found. Rodeo is big business, east and west. The art of pleasure driving, almost esoteric a few years ago, is in spirited revival. From pickup games to full-fledged international handicap tournaments, polo continues to renew itself, with a fine disregard for the inroads of property, income, and estate taxes. The 4-H Clubs and United States Pony Clubs instruct, inspire, and encourage children and their backyard mounts in the pursuit of excellence and companionship, while groups dedicated to horsemanship for the handicapped provide therapeutic riding instruction for physically, mentally, and emotionally impaired children and adults.

The citizen-sponsored American Horse Council (partially underwritten by federal funding and indicating by its very existence the importance of the horse to the national economy) has stated that there are now more horses in this country than there have been at any time in our history. Every one of these animals, having been bred here or imported, must be housed, fed, tended, shod, doctored, trained, bought, or sold. Performance animals must be supplied with suitable tack or harness, and perhaps with a vehicle, with transportation as necessary, and with riders or drivers (in appropriate turnout), who it is hoped have received proper and painstaking instruction.

The underlying reasons for this newly flourishing horse industry are conjectural, but surely the equine renascence derives in some part from the ages when horsepower was no metaphor. To own a horse conveys, perhaps, a certain status; to mount one literally raises the rider above the crowd; to control, by word or touch, this mighty, flighty creature is an achievement. Moreover, horses recall us to a less frenetic time and compel us to work against a slower

6

"Horse Racing" (from The Gentleman's Recreation, *published in England by Richard Blome in 1686)* *illustrates the most venerable of horse sports.*

clock. They gestate slowly, mature slowly, and may be successfully taught only by deliberate, patient degrees. They require a regulated regimen and must not be hurried over meals or digestion. Their delicate and innocent self-absorption is soothing to our fragmented selves.

The horse provides a gymnastic challenge to the athlete, an exercise in calculation to the gambler, a vicarious thrill to the sports spectator, a symbol of power to the successful, an element of pageantry to a procession, a grace note to a rural view. In fact, horses have been a part of human history for so long that apparently we simply will not do without them.

When people build permanent dwellings, they also build more or less permanent shelters for their livestock, and for the same primary reasons: that is, for protection from the elements and from predators, and for the convenience of owners or attendants. As it was in the beginning, so it is today. In stable buildings, form has consistently followed function; typically, all animal housing exemplifies the Mies dictum ''less is more'' and often achieves a certain elegant distinction thereby.

There are, of course, many triumphant exceptions to this rule: stables which have been intentionally endowed with striking architectural qualities unrelated to their purpose. As essential servants, horses were kept convenient to hand, so that the stable often forms part of a total visual scheme, but, in addition, stables have been built to reflect the wealth, position, or pretensions of their owners. Horsepeople, no matter how venturesome or even uninhibited

''The Godolphin Barb, Ancestor of the Most Celebrated Racers on the British Turf'' and one of the foundation sires of the Thoroughbred, receiving the stable cat in his palatial loose-box.

8

they may be in other respects, are conservative about the care of their animals, disliking change only a trifle less than those dedicated creatures of habit, the horses themselves. Yet there is still room for considerable variety in stabling, inside as well as out.

A stable may have straight stalls, open at the back and just wide enough to lie down in, in which horses are tied side by side, separated by partitions of optional heights; or it may have loose-box stalls in which a horse moves about more or less freely; or it may have both. Straight, or tie, stalls must have an aisle behind them wide enough for horses and people to come and go, and, unless their floors are clay or dirt, they will generally rake or pitch to the aisle for drainage. Box stalls may open onto the aisle, or to the outside, or both; their doors may be solid or Dutch, sliding or hinged, with or without barred and meshed panels, and centered on the module or set in the corner. The boxes may be square or oblong, with square or beveled corners, and they range from a minimal nine feet by nine to a palatial sixteen-foot square (or even more, as in a foaling stall). They may be provided with projecting "kickboards," placed about two feet above the floor in order to prevent a horse from being cast, unable to rise. There may also be "tailboard" buffers at wainscot height to avoid damage to high-set tails. Stalls may have built-in tie rings, mangers, hayracks, waterers, salt blocks, heat lamps, foal pens—or none of the above. Aisle and stall floors may be dirt, clay, wood, sand, quarry dust, brick, asphalt, cement, modern compounds such as rubber or pressed cork, or synthetics such as Tartan. Walls and partitions may be metal, wood, brick, cement block, or stone, and may or may not have slatted, barred, or meshed openings between them. Every horse person can and probably will take a firm position on the merits of these alternatives. But a certain consensus can be reached. A good horse barn must be light, and airy without being drafty. There must be handy storage for feed, bedding, and equipment, and there should be a separate space in which to store tack, harness, and vehicles.

A stable is a rest area for horses, and should conduce to their comfort and relaxation, but for people it is a work area, and ought to be arranged to facilitate human tasks. Within this framework the permutations, combinations, and refinements are almost infinite, but there have been no profound changes in the principles of good stabling for centuries. Save for the increased use of the box stall, most really noticeable changes have come about in our busy mechanized century, and in response to advancing technology and an ebbing labor pool, rather than from any great instinct to progress.

Within living memory, hay and bedding were almost universally stored overhead, ease of distribution outweighing the risk of fire, and horses were led to water as part of their daily schedule; one waited patiently until they saw fit to drink. Haylofts are still built, of course, and water hauled in buckets. Horse-keepers still trudge back and forth from stable to manure pile (or pit). It is, however, perfectly practicable to store inflammable hay and bedding in a separate building, to pipe water to every stall, and even to remove manure and soiled bedding via a recessed conveyor belt running through the rear of the loose-box range. This last refinement is built on the same principle as the continuous chain of dump carts still seen on ore tips. Unfortunately, in a stable application, unless it is loaded very neatly it will jam. A good deal of Yankee ingenuity has been applied to speeding up and simplifying stable work, but the wheelbarrow, the manure fork, and the broom are far from obsolete.

Because contemporary building and maintenance costs generally dictate that available funds be applied to practical rather than aesthetic ends, new stables of real architectural ambition are rare. They do, however, exist. In a commercial installation, a striking background can be justified as advertising; an expensive and elegant building suggests that the horses within must be of commensurate worth. But there is a deeper rationale behind any apparent extravagance—the common feeling that one's horses should live well. Equine appreciation of scale, mass, and ornamentation is undoubtedly limited, but if such values are important to the owner, he will do his best to see that his horses are no more stinted in this respect than in any other. Stables still tend to reflect the priorities as well as the resources of those who commission them.

In the eighteenth and nineteenth centuries that part of America which now constitutes the continental United States was pretty thinly peopled. The population did not reach fifty million until about 1880, and even then more than half of it was east of the Mississippi. Most people lived out their lives without traveling more than a few miles from the place where they were born or had settled, and accepted modes of living changed very little from generation to generation. Society remained highly stratified according to inherited European patterns, although this stratification came more and more to reflect comparative financial position than any feudal standard. Colonists and early immigrants retained their inherited conceptions of their own and others' social status, and although they commenced to make the transition from class to class with considerable celerity, class distinctions themselves were slow to blur. In architecture as in other things, the farmer and artisan and the aristocrat or landed gentleman had somewhat different aspirations. In their outbuildings as in their houses both strove to reproduce, with suitable modifications in scale and materials, the sort of design schemes that their ancestors had known. These twin threads of economical simplicity and deliberate grandeur, their origins half-forgotten, continue to weave their patterns through stable design to this day.

II *A Goodly Heritage*

. . . and laid him in a manger, because there was no room for them in the inn. —LUKE 2:7

Illuminated manuscripts have survived the ravages of time astonishingly well. By setting their vignettes in familiar landscapes, the medieval miniaturists who embellished these documents have left us an unequaled research source on the details of dress and architecture in their times. In the glowing Nativity scenes from fourteenth- and fifteenth-century missals and books of hours, the aesthetic and pragmatic roots of even the most modern stable design may be found. Again and again these inspired and meticulous craftsmen have recorded some long-vanished humble stable with touching fidelity. The little frame buildings are curiously familiar in appearance, in spite of their thatched roofs and wattled partitions or walls. Sometimes, of course, the artist has idealized the scene. Where the Virgin's gown is furred and the Babe wears a crown or lace-trimmed robe, we may perhaps see only a detail of the stable: a marble arch or part of a noble cornice, only a suggestion of the size and majesty of the imposing pile outside the picture.

The great patrons and architects of the Renaissance endowed their stables with a magnificence appropriate to the position of the owners and the caliber of the inhabitants. For example, the Lipizzan horses of the Spanish Riding School in Vienna have been housed in the Stallburg Palace, behind the lowest tier of the grand arcaded facade, since 1746. Royal indeed, their lodgings, gleaming with brass and marble, have been exhaustively described and photographed. Not the least awesome feature of these stalls is that they have been in continuous use since they were completed for the horses of the Imperial Guard in 1565, and are probably the earliest stable forming part of a stately residence to have survived into our time.

In England, the first Duke of Newcastle's great stable at Welbeck has long since disappeared. William Cavendish (1592–1676), the builder, was a man of

11

The Adoration of the Magi,
*from the fifteenth-century
Llangattock Hours of the
Virgin, decorated in the school
of Jan Van Eyck and Petrus
Christus.*

many talents, enormous wealth, and extensive properties. He was also a passionate and creative horseman, and, like most horsemen, a man of firm and sometimes controversial opinions on equine training and management. He served in the governments of both Charles I and Charles II, and spent sixteen expatriate years in Belgium for supporting the Royal cause during Cromwell's rebellion. In Antwerp he conducted a riding school and training establishment, and published there in French the first (1657) edition of his seminal work, *A General System of Horsemanship in All Its Branches*, complete with handsome and detailed plates by Abraham Diepenbeke to illustrate his precepts.

12

COLOR PLATE 4. *Entrance front, Russborough, Co. Wicklow. The stable is on the left.*

COLOR PLATE 5. *Wadsworth stable, Lebanon, Connecticut.*

COLOR PLATE 6. *Domestic stable and carriage house, Mount Airy, Virginia.*

In these drawings, perhaps out of nostalgia, the noble exile had his model stallions posed in front of prospects of his estates. Several of these are views of Welbeck, the principal seat of the Cavendish family, and of the Duke's stables and Riding Hall, designed for him by John Smithson in the 1620s. Of these structures only the exterior of the Riding Hall survives. Its top-lit open arena, measuring forty by one hundred and twenty feet, and its graceful hammer-beam roof were swallowed up in 1889 when the interior was converted into a library and chapel. The stable buildings, which included commodious apartments for the stud manager, grooms' quarters, and a large granary, were pulled down in the 1750s, but we know a great deal about them from contemporary documents.

William Cavendish's great stable at Welbeck Abbey, Nottinghamshire, as engraved by Diepenbeke, ca. 1657.

The exterior architecture was magniloquent, displaying Smithson's curious composite pediments and oriental towers, together with clustered chimney pots and cross-mullioned Tudor windows. The interiors were of stone, with flagged floors and vaulted ceilings. There were fifteen straight stalls in each row, each measuring five and a half by nine feet, with a generous aisle behind them. A marble manger ran the length of the range and conducted a continuous stream of fresh water through the stalls. The windows were high up in the thick and lofty walls, but at head level in each stall was a small shuttered tunnel-vent, which could be opened or closed according to the wind and temperature. The upper stories, above the fireproof ceiling, appear to have been used for storage, but may have provided additional grooms' quarters; there were at least four flues serving one stable block. We should note that the Duke insisted on lavish bedding, and was extremely particular about his feeding systems. His run-in shelters and feed racks have of course long since disappeared, but from Diepenbeke's engravings we may see how little such designs have changed.

Detail from Diepenbeke's engraving of Welbeck animals at pasture.

In 1734 Welbeck passed, by marriage, into the Portland family, but the Cavendish predilections for building and for horses were dominant traits, finding their fullest expression in the latter half of the nineteenth century with William Bentinck, the fifth Duke of Portland. Under his aegis Welbeck "grew to enormous and palatial proportions." Besides his additions to the Abbey proper, he built a new Riding House, three hundred and ninety-six feet in length by one hundred and eight feet in width, and rising to a height of fifty feet. It had a glass roof modeled after the roof of the old Riding House, and additionally was lit by four thousand gas jets. It was, in its time, the largest indoor school in the world, save for one in Moscow. The Duke also built new stables, on the still common quadrangular pattern, in a succession of independent yards; his hunting stables alone covered an acre of ground. A bachelor, he was succeeded in 1879 by another Cavendish descendant. The sixth Duke twice served as Master of the Horse in Queen Victoria's household and enjoyed considerable success on the Turf (although, under the "pervasive influence" of his Duchess, he gave up his racing interests and sold his Stud).

Even the later splendors of the Welbeck establishment are put in the shade by the most magnificent of all the Renaissance stables, designed by Jules Mansart and built between 1679 and 1682 for Louis XIV at Versailles. The Palace windows look out on massive twin horseshoe facades, separated by the Paris road where it runs into the Place d'Armes. The Grande and Petite Ecuries are actually the same size, but were so designated because the Grande Ecurie housed the *chevaux à main,* or riding horses, while the Petite Ecurie sheltered the coach horses and their vehicles. In addition to a huge staff, the King's pages and sons of the nobility educated at court, together with their instructors, were billeted in the stable blocks. Here also they had their lessons, in which equitation played a major part. The pages' chapel was in the buildings, and a small prison; the enormous covered manège of the Grande Ecurie has also served as an opera house and ballroom.

The horses, nevertheless, were of primary importance, and were accommodated accordingly; the Elector of Hanover remarked that the horses at Versailles were better housed than he was himself in Germany. They were certainly a valuable lot, and must have formed the most cosmopolitan equine community in history. After 1663, no French horse could be sold without prior notice to the King's agents, who had preemptive rights. Any economies this edict effected, however, were more than outweighed by the enormous sums paid for imported stock, which was brought from England, Denmark, Spain, Italy, Morocco, Algeria, and Persia. The Royal Stud carried on an extensive breeding program as well; the Anglo-Norman strain (forerunner of the French Saddle Horse) was developed at Versailles. When Louis XVI lost his throne there were more than thirteen hundred horses in the Ecuries.

The Revolution put an abrupt end to this regal extravagance. The animals were dispersed, and most of the vehicles and equipment sold — or looted. Until

16

well into the twentieth century these paired masterpieces of Mansart's design, neoclassic in conception, baroque in ornamentation, were put to various mundane uses, or simply stood vacant. Lately, the Musée de l'Oeuvre, a museum of the history of Versailles, has been established in the Grande Ecurie amid royal equine shades.

The ambitious stables at Welbeck are gone, those at Versailles remain as monuments to past glories, but at Badminton House, in Gloucestershire, the far more conservative but superb complex begun about 1658 by Henry Somerset, Lord Lieutenant of Gloucestershire, first Duke of Beaufort, and Lord Lieutenant of Wales, still efficiently serves its intended purpose. Although much altered, expanded, and improved in the eighteenth and nineteenth centuries, these buildings adhere to their original austere simplicity; their plan and style have had a recurrent influence on later designs.

Opposite: The Sun King's stables at Versailles, by Jules Mansart, from a 1690 painting by Jean Baptiste Martin.

The Kips engraving of Badminton House, Gloucestershire, and its environs, ca. 1660. The stable is to the right of the court.

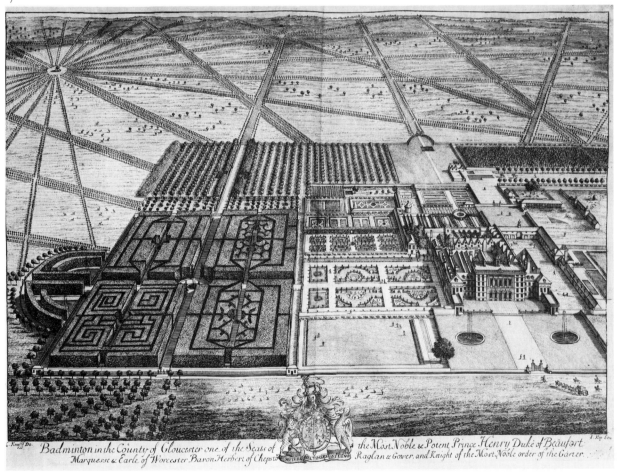

Badminton in the County of Gloucester one of the Seats of the Most Noble & Potent Prince Henry Duke of Beaufort Marquesse & Earle of Worcester Baron Herbert of Chepsto Raglan & Gower, and Knight of the Most Noble order of the Garter.

The first Duke lived at Badminton in great state, sheltering nearly two hundred persons in the "large expanded house" which eventually evolved into the present grand Palladian composition. He nevertheless kept a keen eye on the practical management of his household and estates. The forty-stall stable he provided for the horses of his guests, adjacent to the visiting grooms' quarters by the carriage entrance to the mansion, established the form of the stableyard as it is today. A contemporary visitor described the original stables as "pompous." He may have meant to imply only that they were comparable to the remainder of the imposing establishment, or it may be that their lines were simplified in later rebuilding and expansion. At any rate, the Kips engraving shows

The stableyard at Badminton House in 1976.

18

the same sort of lofted single story, surmounted by plain double pitched roofs, broken once in each side of the square by a central triangular gable. Today the wide, well-lit brick aisles give ready access to single rows of roomy loose-boxes under a handsome vaulted ceiling. Feed and tack rooms are conveniently situated. The quadrangular plan conceals the workings of the stable from the residence and from visitors arriving at the carriage entrance, but it is a much shorter walk from that door to the stable than it is from one end of Badminton House to the other. Competitors at the Badminton Trials (the oldest and perhaps best-known three-day Combined Training Event in England) use these stalls, which have housed at one time or another most of the greatest Event horses in the world, but for two-thirds of the year they are occupied by the horses of the Badminton hunt.

The first Duke, who bred all his own horses for domestic and estate use, and kept a pack of hounds, was upholding a tenacious family tradition of hunting and horsemanship that is still unbroken. The Badminton, under the uninterrupted Mastership of the Dukes of Beaufort, is the oldest established pack in England. In addition to a succession of notable hunters and coach horses, the Ducal stud farm has produced racehorses of renown. That noted whip and foxhunter, the eighth Duke, served as Master of the Horse to Queen Victoria and was the patron of the famed Badminton Library of Sports and Pastimes; the tenth Duke assumed the office of Master of the Horse in 1936, retiring only in 1979, and has been Master of the Beaufort Hounds for over fifty years.

The ideal residential stable, of course, is convenient of access to owner and staff, contributes to the aesthetic value of the establishment, and at the same time takes advantage of the site to provide the healthiest exposures for its tenants. An ingenious solution to this problem was supplied by Andrea Palladio, the sixteenth-century North Italian architect whose surviving works, largely in and around Vicenza, demonstrate his advocacy of the balanced architectural principles of Roman antiquity. The convention, identified with his name, of linking outlying wings to a central block by colonnades or curtain walls, offers many advantages. The half embrace of the often curving colonnades has a welcoming air and draws the eye to the center of the composition, while the wings (which may well conceal utilitarian farm buildings behind their elegant facades) serve to extend an already imposing front and at the same time keep offices and stables close at hand, a defensive medieval convention surviving in many a later scheme.

Palladio studied and measured the remains of various ancient buildings, and in 1570 published *The Four Books of Architecture*, which contain his conclusions as well as his own architectural designs to that date. The work became a pattern book for a purified classical movement. Inigo Jones (1573–1652) was the herald of this style in England; its chief exponents there were Lord Burlington (1694–1753) and Colen Campbell (died 1720).

Castletown House, Co. Kildare: the entrance front. The stable is at the end of the arcade to the right, balancing the kitchen wing.

Palladianism, particularly with respect to country houses, reached Ireland early in the eighteenth century. Castletown, Co. Kildare, is the earliest of the Irish great houses drawn to such a plan. Kitchen and farm offices occupy the west wing; the east wing contains the stable. Begun in 1722 by Thomas Conolly, Speaker of the Irish House of Commons and by all accounts the richest man in Ireland, Castletown was built largely to the design of the Italian architect Alessandro Galilei (1691–1737), with considerable contributions by the owner, as well as by others whom Bishop Berkeley, himself a student of Palladio's works, tartly described as "many heads."

The interiors of the house itself remained incomplete until after 1758, when William Conolly, the heir, married Lady Louisa Lennox. She immediately set about the finishing, and for sixty years continued her alterations and improvements. The house has hardly been changed since her death in 1821, save for the conversion of the grooms' quarters above the stable into elaborate bachelor accommodations fifty years later. Castletown remained in the Conolly family until

20

1965. It is now the headquarters of the Irish Georgian Society, and open to the public.

The stable is in an excellent state of preservation. The floors are brick throughout, the raked stall floors raised above the level of the aisle, with their bricks laid in the same running bond as the aisle, but perpendicular to it. Mangers, hayracks, and other surviving hardware are of a plain, practical pattern. The partitions separating the generous straight stalls curve gently from the wall to the round stone columns of classically correct proportions that support the vaulted ceiling. The whole is well lighted, airy, workmanlike, and happily once more in use; hunters and polo ponies are boarded there.

Interior view of the stable at Castletown.

Of the army of enormous country palaces created in Ireland during the eighteenth century, a number are in the Palladian style, and several of these, including Carton, Castletown Cox, and Kilshannig, incorporate stables into their composition. The German-born architect Richard Cassels (Castle) was responsible for many of these designs, of which Russborough (1742), Co. Wicklow, is his chef d'oeuvre. At Russborough, the front is extended beyond the wings by twin pavilions connected to the flankers by walled courtyards with arched central gateways. The west pavilion was the stable, but as the length of the composition is seven hundred feet overall, the horses were still a pretty long way from the house.

The entrance front of Russborough, Co. Wicklow. See also color plate 4.

Carton House, Co. Kildare; the stable is to the left.

22

III Early Efforts

*Send [with the colonists] . . . horses and mares, and all other things
necessary for said plantation.* —JAMES I, VIRGINIA CHARTER (1609)

Castletown and its fellows were soon to have their counterparts among the established plantations of the Virginia Colony, but the Palladian was not the first design scheme for a self-sufficient establishment to travel first to Ireland and then to America. When the earliest colonists from England were dispatched here at the beginning of the seventeenth century, under charter from the London Company, with them went the detailed plans for setting up a pioneer community that were in current use in Ulster: plans drawn up with a view to protecting settlers and their livestock from hostile natives by enclosing all essential support systems within a bawn, or walled enclosure.

The primitive outbuildings of these daring people have vanished along with their homes, but, surprisingly enough, not without trace. At Carter's Grove, James City County, Virginia, an archaeological team seeking remains of the eighteenth-century supporting buildings of that great Georgian mansion has lately discovered the site of the ill-fated colony of Wolstoneholme, which was burned and abandoned at the time of the 1622 massacre instigated by Chief Opecancanough. Archaeologists are also investigating a similar community, which met the same fate, on the other side of the James River. Conjectural reconstruction of the tiny settlements at Wolstoneholme and Flowerdew Hundred show that these British outposts, the earliest yet identified on these shores, approximate rather exactly the pattern of the Irish bawn villages. The first domestic building to be erected was apparently a community longhouse within the palisade, with an adjacent shelter for animals under the same thatched roof.

The first horses shipped to Jamestown were eaten at the time of the great famine, but more followed: in 1620 the London Company sent "twenty mares, beautiful and full of courage," doubtless to a stallion already available. The

23

modest shed at Flowerdew Hundred, an even humbler version of the little building in the Llangattock *Adoration* (page 12), may have lodged only a cow, a few sheep, or even a sow and her litter. But horseshoes and spurs have been found there, and a sixteenth-century stirrup was turned up at Wolstoneholme. It is tempting to think that these shadowy postholes in the yellow clay of the James delta mark the sites of the first American stables. Certainly horses soon began to play their part in the commerce and communication of the colonies.

Long before the Revolution, when roads were few, ill made, and almost wholly unmaintained, two new equine breeds were developed to meet the exigencies of overland travel; both disappeared quite early in the nineteenth century, as trails and paths gave way to graded roads. The Narragansett Pacer was the first of these, descended from English and Dutch stock brought to the Massachusetts Bay Colony by the early settlers, and refined by the racing aficionados of liberal-minded Rhode Island. This small, short-necked, sturdy, and rather ugly animal, with its heavy mane and crooked hind legs, was superbly adapted for passage through an untamed country. Dr. James McSparran, an Irish clergyman resident in Rhode Island, who made extensive journeys in the course of his duties, even as far as Virginia, stated that sixty or seventy miles a day was not an unusual distance for a Narragansett to cover in its deceptive "rocking chair" gait. The pace is, on balance, a faster gait than the trot. Although Narragansetts rarely measured much over fourteen hands, Dr. McSparran (surely a credible witness) reports that some of them could pace their mile in not much more than two minutes.

With the increase in vehicular travel, the Narragansett lost its popularity and died out, superseded by bigger and better-looking animals carrying Thoroughbred and Arab genes, and more suited to carriage work. But its much diluted blood still flows in Morgan, Saddlebred, Standardbred, and Paso Fino lines, even though we cannot show a stable where we know the animal was housed. It is claimed that Paul Revere borrowed a swift Narragansett on that famous midnight, but Dean Larkin's barn has vanished along with the tough little horses he kept there.

In early eighteenth-century Pennsylvania, the German settlers used oxen for heavy draft work, but by 1750 some Belgian and Flemish horses had been imported, and by crossing these on lighter breeds the practical farmers evolved the Conestoga horse. Clean-legged, quick but powerful, standing between sixteen and seventeen hands and weighing close to a ton, the Conestoga supplanted the ox on Pennsylvania farms for fifty years, and for the same period hauled heavy freight wagons over the Alleghenies. Conestogas were probably the rightful tenants of the stone barn at Valley Forge that sheltered General Washington's horses, but it was not long before the breed gave way to slower but stronger purebred strains, as roads improved and larger and heavier Conestoga wagons were built.

24

Iron spur, evidence of the presence of horses at Flowerdew Hundred, one of the earliest European settlements along the James River.

The farm stable at Valley Forge, where Washington's chargers wintered. The reconstruction of the interior is conjectural.

Reconstruction of the framework of an early seventeenth-century New England all-purpose barn, photographed during the re-creation of Plimoth Plantation, Plymouth Massachusetts.

Left: Log stable at Connor Prairie Settlement, Noblesville, Indiana, ca. 1826. Similar barns of a somewhat later vintage may still be seen in remote districts. "Snake" fencing followed the frontier to the edge of the treeless plains.

Right: The tule shelter at the Hugo Reid Adobe, Los Angeles State and County Arboretum; a reconstruction of the 1849 original. See also page 168.

After grievous early struggles, the colonists of the eastern seaboard rapidly established themselves in as civilized and orderly a society as the shadow of the wilderness allowed, and soon began to put up domestic buildings of an admirable solidity and considerable intrinsic beauty. No practical or aesthetic considerations dictated the preservation of the rough shelters for man and beast thrown up by the earliest arrivals. On the seaboard they were replaced as soon as suitable labor, equipment, and materials became available, but they continued to appear, if in more efficient form, along the fringes of European settlement until well into the nineteenth century.

The actual embarrassment of timber made available by clearing land made log construction, first introduced by Northern European settlers in Delaware, extremely practical, and the technique of laying up such a building became very well understood. There are eighteenth- and nineteenth-century log houses, chinked or weatherboarded, still in use today, and their practical advantages, combined with their appeal to our new preoccupation with historical and environmental concerns, have inspired a revival in their design and construction.

26

There remain, however, very few old log barns. The log stable shown here is part of the exhibit at Connor Prairie Pioneer Settlement, an historical living museum project of Earlham College, which portrays the folk life of central Indiana in the 1830s. The building is largely original, although it was moved to its present site. It is the most primitive permanent structure in the hamlet — a survival of what was even then a vanishing era.

Colonial stabling on the West Coast was more heterogeneous, and much of it is, of course, of a later date than the earliest examples in the East. In what is now southern California, where the first serious Spanish colonization began after the middle of the eighteenth century under the inspired leadership of Father Junipero Serra, missions, public buildings, and homes were alike built of adobe brick and roofed with tile or tule thatch, laid over willow rafters. In that benign climate the comfort and safety of domestic animals may be ensured by putting up a ramada, a roof of loosely woven tule reeds (more likely nowadays to be plastic stripping) supported on willow posts and surrounded by a fenced corral. The matting filters the sun and sheds heavy rain, while taking advantage of every vagrant breeze. This extremely well-adapted shelter is still, in various guises, in common use, although frequently in conjunction with more conventional stabling.

We have been unable to discover any extant California stable building proper earlier than the 1844 stable at Sutter's Fort, reconstructed in the 1940s.

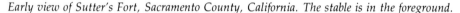

Early view of Sutter's Fort, Sacramento County, California. The stable is in the foreground.

27

After California achieved statehood (or had it thrust upon her) and gold was discovered in the Sierra, a construction boom began that has, perhaps unfortunately, continued to the present day. It is fair to say that California, settled variously and competitively by Spaniards, Russians, and Easterners of various European stocks, can show homes and outbuildings in every taste. But the heavy-walled cool adobe and the open tule shelter, no matter how altered or enlarged upon, still suit the landscape and the climate.

Detail of an early lithograph of Lachryma Montis, the 1851 residence of General Mariano Vallejo, Sonoma, California. The restored masonry and frame stable is now a reception center.

Left: The big bank barn at Crebilly Farm, eastern Pennsylvania. Built for farm horses, it now houses cattle, in a reversal of the usual evolution.

Right: A typical eastern bank barn with byre and stable below, erected on the California coast in the 1880s by a transplanted New Englander.

North of New York, on the East Coast, strong Puritan tradition and harsh winters made snug homes and outbuildings the standard; although often handsome as well as utilitarian, they were generally less expansive than their counterparts below the Delaware. In the New York area and across New Jersey into eastern Pennsylvania and central Ohio, the early Dutch and German settlers produced solid stone barns and stables, often building them into a slope so that upper and lower floors could be entered at ground level.

These bank barns, set on massive stone foundations and bearing walls, blaze a path of western expansion across the continent. The upper stories may be stone, brick, or frame, but the solid granite underpinnings are a signature that recurs in the Valley of Virginia, in the Bluegrass region of Kentucky, and at last, toward the end of the nineteenth century, on the West Coast. They mark the homesteads of the lovers of the soil, the true farmers who, in the wake of the dashing fur trader, buffalo hunter, and prospector, drew their sustenance and modest prosperity from the land itself. Some of these barns were all-purpose buildings, some were always horse barns; a substantial number were built for cattle and have been converted into stables.

The mid-nineteenth-century five-story Shaker stone barn, West Lebanon, New York.

Only the walls of the West Lebanon barn, burned out in the 1970s, still stand.

30

In New England, the sensible convention of appending woodshed, stable, and minor subsidiary buildings to the residence in an interconnected range was established early, but the same climatic conditions that dictated this precautionary layout also led to the construction of some really enormous stone or frame barns, reminiscent of European tithe barns, which held all the livestock, vehicles, farm gear, and stored crops under one roof. Their noble proportions, structural integrity, and efficient use of space are impressive and highly attractive to pen and camera, although stabling played only a modest part in their grand scheme.

Stables and coach (or carriage) houses were at first more common in urban areas than in the countryside, and were often given a finish to complement the houses they supported. In 1730, the Reverend Daniel Wadsworth built an elegant mansion and stable in Hartford, Connecticut (color plate 5). His son Jeremiah succeeded to the property, and as Commissary General of the Continental Army, entertained most of the luminaries of the Revolution there. The best stalls in the spacious frame stable were, of course, reserved for General Wash-

Garth's Auction Barn, an old stone and frame building near Columbus, Ohio. Note the Dutch stall doors under the projecting loft.

The Derby Farm barn, designed by Samuel McIntire (see page 51), now part of the museum complex at the Abraham Browne House in Watertown, Massachusetts. The tea house has been moved to Danvers, Massachusetts.

ington's horses. His favorite mount, Nelson, enjoyed the questionable luxury of the single loose-box, which is little more than one of the tie stalls closed in, but does provide the comfort of a level floor. There are two straight-stall ranges in the carefully preserved interior, one with outside ventilation and the other butting the partition to the carriage room across a wide aisle space. A hayloft and possibly coachman's quarters occupied the upper story; above the vertical hayracks there are drops, which also serve to exhaust hot air by way of the cupola in the roof. The stalls themselves are very slightly pitched and have double floors of heavy planking, which appears to be elm.

The building itself is a simple square, with shallow gable roof over a low upper story. The exterior, however, has been endowed with an elegant dressing of applied classical detail. The roof line has been disguised by false fronts at the

32

Stable and carriage house at the Rundlet-May House, Portsmouth, New Hampshire, ca. 1807.

gable ends, segmenting the roof in such a way as to produce the effect of a tripartite building; the central block is capped by a severe pediment and flanked by wings with great arched openings. The pseudo-pediments rest upon wide applied pilasters, employing pronounced entasis, which are repeated on all four sides of the building; and false carriage doors and arches in the rear echo the openings in the front. Only the windows and grooms' doors, prosaically cut where they are most useful, flaw the painstaking symmetry of the effect.

Threatened by an urban renewal scheme, the Wadsworth stable was acquired by the Daughters of the American Revolution and moved in 1954 from Hartford to Lebanon, Connecticut. It appears oddly at home in its bucolic setting, where it may be visited by arrangement with the Curator of the Jonathan Trumbull House, next to which it stands.

The unknown architect's projected elevation for Drayton Hall, near Charleston.

The Palladian touches on the Wadsworth stable are entirely decorative, lending imaginative grace to what is otherwise a rather clumsy box. On the plantations of the South, Palladio's influence was felt in principle as well as in embellishment, as may be seen at Drayton Hall, which is probably the finest example of our early Georgian architecture, and the lone survivor of the great plantation houses on the Ashley River, in the Low Country of South Carolina. Built in 1738, it remained in the Drayton family until 1974, when it passed into the hands of the National Trust for Historic Preservation. Only the glorious central block remains, but the unknown architect's drawing shows the full projected composition of house, supporting offices, and attached farm buildings. Although the flankers were taken down in this century, and the ambitious stable wings were never built, the plan was in accordance with Palladio's precepts on the organization of a country estate, placing the farm buildings ". . . not too near the master's house . . . nor so far off as to be out of sight." The envisioned "economical Palladian layout" of Drayton Hall is an American echo of Castletown or Russborough.

Eighteenth-century Virginia agriculturists, it was said, built their barns to shelter their crops, not their livestock, a custom described as "pernicious" by a foreign observer, although it is one still successfully adhered to on western ranches. Even then, however, Virginia took a rightful pride in her hunters, carriage horses, and racehorses, and stabled them with appropriate care.

34

One of the earliest of these domestic stables is at Shirley Plantation, on the James River in Charles County, where the whole complex constitutes a special survival. The original nine thousand acres was patented by Edward Hill in 1660; the buildings date from before 1740 and are the work of John Carter, who married Elizabeth Hill in 1723. This working plantation, still in the Hill–Carter family, is open to the public all year.

The approach to Shirley from the land side (the formal entrance of course faced the river) is along a modern poplar avenue, opening into a forecourt formed by two ell-shaped storage barns and twin two-story dependencies closer to the house. Stable, dovecote, and smokehouse are to the left of this group. Like the rest of the outbuildings, the stable is brick, laid up in Flemish bond. The north gable was built with a jerkin-head, which was removed in the 1880s or 1890s to enlarge the loft space; the ceiling joists were lowered at the same time, and it appears that the central carriage arch was partially bricked up into its present form. The sill is above ground level, the original approach having apparently been worn away. The interior floors were brick laid in sand, and the horses occupied tie stalls with high hayracks. The remaining stalls still house the riding horses of the Carter family and their guests.

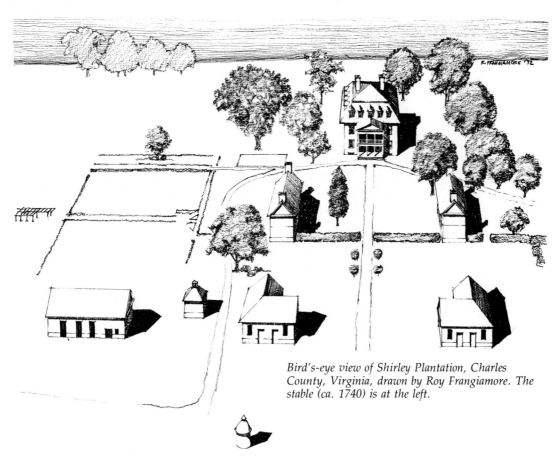

Bird's-eye view of Shirley Plantation, Charles County, Virginia, drawn by Roy Frangiamore. The stable (ca. 1740) is at the left.

The dependencies which are freestanding at Shirley are combined at Mount Airy, in Richmond County, into the first complete Palladian composition in the country. Mount Airy has been the seat of the Tayloe family since the first Colonel John Tayloe, drawing his inspiration from the published drawings of Robert Adam and William Gibbs, completed it in 1758. It is one of the most beautiful houses in America, and its builder and his descendants played a considerable part in the history of the American Thoroughbred. Colonel Tayloe owned the American stallion Yorick, which at the age of thirteen, and after six seasons at stud, won a five-mile match race in twelve minutes and twenty-three seconds, carrying one hundred and eighty pounds. But more impressive records were to come. In 1759, the Reverend Andrew Burnaby, visiting the Colony, wrote, "The Gentlemen of Virginia, who are exceedingly fond of horseracing, have spared no expense or trouble to improve the breed of them by importing great numbers from England." These sporting gentlemen included the Tayloes; and

The Pendleton lithograph of Mount Airy, Warsaw, Virginia. See also color plate 6.

36

Mount Airy's greatest glory was in the time of John Tayloe III (1771–1838). Between 1791 and 1806 the Tayloe horses made one hundred and forty-one starts, winning one hundred and thirteen of them. The first astonishing success of the Mount Airy stud was brought by four stallions by the gray horse *Medley,† who passed on his color as well as his speed and soundly demonstrated the fallacy of the intermittent old prejudice against grays (which is really only admissible in overworked and disgruntled grooms). In 1798 Colonel Tayloe took a half interest in *Diomed, just out from England at the advanced age of twenty-one. From *Diomed he bred, out of his blind mare *Castianira, Sir Archie, one of the great foundation sires of the American Thoroughbred, and at a remove, of the American Standardbred as well. Sir Archie stood for most of his career in North Carolina, but sired for Mount Airy such stars as Lady Lightfoot (who was foaled at Belair, in Maryland; see page 112). Sir Archie's influence on American bloodstock can hardly be overstated, although it was many generations before the blindness that afflicted his sire and dam ceased to reappear, notably in the intransigent but prepotent Boston, famous sire of both trotting horses and Thoroughbreds, and in Boston's son Lexington, who up to that time was certainly the greatest racehorse foaled in America, and the leading sire.

The racing stables at Mount Airy have vanished, although the outline of John Tayloe III's training track is visible. The stables were probably frame and have been burned or demolished; we have come across no eighteenth-century training or racing stables at all, nor even any precise description of one, although there were a great many of them. The Tayloes' coach stable, a masonry building with wooden gable ends, probably antedating the mansion, does survive (color plate 6). It holds a central carriage room, with a stall chamber containing six roomy tie stalls to either side; the floors are brick. It is an attractive building, simple but well conceived, and is still in regular use. The six-horse coach teams of an earlier age have given place to the succession of ponies or donkeys of character kept for the pleasure and instruction of the present generation of Tayloe grandchildren; these pleasant creatures contrive to adorn the dignified quarters of their doughty predecessors.

In 1798 Colonel Tayloe built, to the designs of Dr. William Thornton, the elegant Washington residence still known as the Octagon House, and now the property of the American Institute of Architects. The carriage house and stable there was a distinguished essay in the classical style, far more elaborate than its rural cousin, but it has unfortunately been demolished to make way for the A.I.A. headquarters that now shares the double lot with the house. In *Have You Kicked a Building Lately?* Ada Louise Huxtable writes, "One remembers the old brick stables that served as a library . . . and wonders how, and where, their obvious lessons of sympathetic materials and urban relationships were lost."

† An asterisk before a horse's name denotes importation to this country.

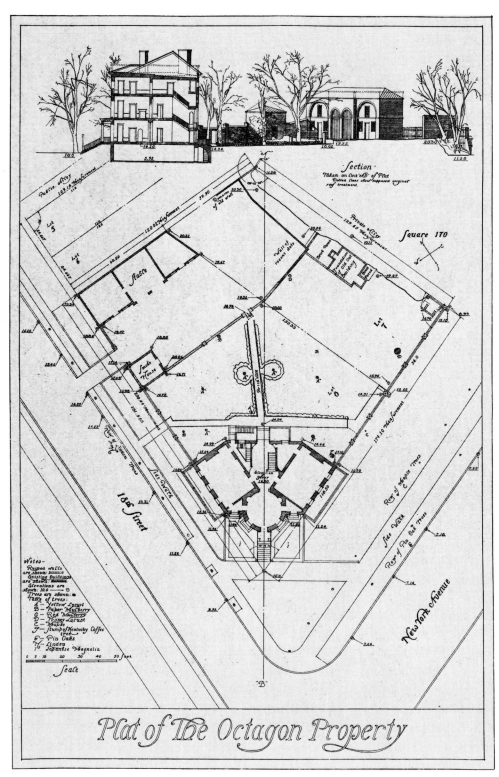

Section

A 1937 architect's drawing of the Octagon House, Washington. The stable is in the left-hand point of the lot.

38

As at Drayton Hall, the outbuildings of Low Country plantation houses have largely disappeared; fire, water, earthquake, and decay have all taken their toll. Even at Middleton Place, famed for its superbly restored and maintained historic gardens and plantation living museum, the great stable in the farm group, its curvilinear gable echoing the design of the preserved wing of the main house, actually dates only from 1930. On the West Branch of the Cooper River, however, one eighteenth-century stable does survive, although in a ruinous state. It was built in 1760, one of a pair of buildings flanking the forecourt of the house at North Chachan Plantation; its twin is gone. The broad roof was fronted by a curvilinear gable with a round central opening, or ocular. Square-headed windows opened to either side of the arched doorway, but the side walls were pierced only by ingenious baffled openings reminiscent of Cavendish's system at Welbeck. As a stable it was also distinguished by an unusual set of mangers, cut into a single cypress log that ran the length of the range. A good deal of brick from this building was taken for the reconstruction of the house, which burned in the 1890s and was rebuilt by a later owner.

Left: Photograph by Frances Benjamin Johnston of the stable at North Chachan Plantation, in the Carolina Low Country.

Right: The stable at Middleton Place, near Charleston.

Tryon's Palace, New Bern, North Carolina. The right-hand pavilion is the stable.

Perhaps the most exciting restoration of an eighteenth-century stable — because of its pivotal importance in the reconstruction of that landmark structure — is that at the Governor's mansion in New Bern, North Carolina. Tryon's Palace, as it was sardonically called, was completed in 1761 from the plans of John Hawkes, English master builder, described as "the first professional architect to remain in America." Naturally he worked in the modish Palladian style. The central block is, in effect, a grand London townhouse, joined by curving colonnades to two-story kitchen and stable wings. It was built as the colonial capitol and Governor's residence of North Carolina, under a truly munificent appropriation by the General Assembly of £75,000. As a footnote to the history of inflation, we may note that the restoration work has cost in excess of $3,000,000.

40

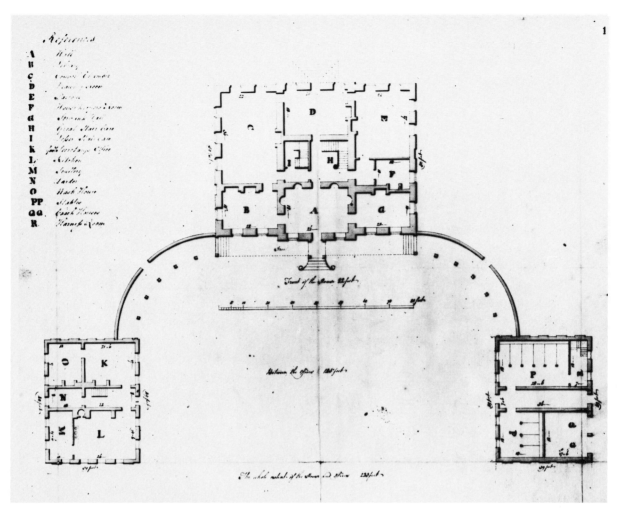

John Hawkes's plan for Tryon's Palace.

The palace stable just prior to restoration. Since the eighteenth century it has served as a schoolhouse, a Masonic lodge, a warehouse, and latterly an apartment house.

After the Revolution, the Palace served as the first state capitol until the seat of government moved to Raleigh. It then fell on hard times, and burned to the ground in 1798. Over the years the kitchen wing and the remains of the house were cannibalized for building materials, and the Palace square disappeared under streets and houses. Somehow, through many adaptations and changes of fortune, the stable block essentially survived, and provided the starting point for the work of restoration in 1952. At that time it was an apartment house, and quite unrecognizable. But because Hawkes's plans and elevations are still on file at the British Public Record Office, and because William Tryon compiled an exhaustive inventory of the contents of the house, the whole restoration may claim an impeccable standard of authenticity; many questions as to brick and bond were answered in the surviving stable walls. The interiors of this building have been as accurately reproduced as those of the mansion. The twin stable areas, with their solid stall partitions and large windows on the aisle, are divided by the carriage room. Fodder was stored behind the blind windows of the second story, which also held rooms for the stable servants. The building is light and airy, the stalls of adequate if not lavish size, with ample brick-floored work space behind them. A curious feature is the want of an interior access between stall and carriage areas; this rather inconvenient arrangement at least kept the Governor's coach from smelling of the stable.

Whitehall, across the Severn from Annapolis, was also a British Governor's residence, designed as a summer pavilion for Governor Horatio Sharpe. It was built by Joseph Horatio Anderson between 1764 and 1768. There is evidence that the prodigiously talented William Buckland also had a hand in the composition, as he certainly did in the embellishment and decoration of the work, but every detail of mansion, park, gardens, and outbuildings received Sharpe's fullest personal attention. Upon his replacement in 1769, the ex-Governor retired happily to Whitehall, which he continued to improve and enjoy until his last visit to England in 1773, from which the Revolution prevented his return. He had never intended to leave his dear Palladian "garden house" for good, and would certainly have come back to it at the conclusion of hostilities had his age and health allowed. At Whitehall he had been free to indulge his passion for gardening and to devote himself to his racing stud, considered even among Virginians and North Carolinians to be one of the finest in the country.

Governor Sharpe was a vigorous and forward-looking man, and must have given much satisfaction to his up-to-date architects. He overcame the difficulty of obtaining suitable stone for the classical facades of his villa by substituting wood (largely hard pine) carved and painted to resemble stone. He installed, in its own polygonal chamber at the west end of the house, a water closet of the modern water-seal type, with marble floors and toilet bowls, and Delft tile walls; this convenience seems to have antedated the earliest known specifications for such a facility by several years, and was the first of its kind in this country. His racing stable, which stood in a paddock northwest of the house,

42

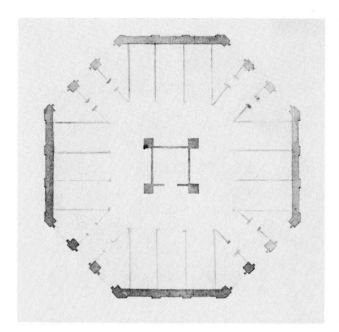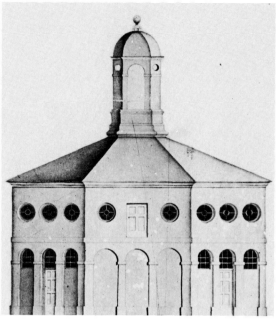

Anderson's plan and elevation for the racing stable at Whitehall, near Annapolis.

was also polygonal, a charming octagon of which Anderson's plan and elevation survive. Its fashionable exterior was probably dictated by its proximity to the house, but, in spite of what current cant termed its "alamodality" and what seems to modern eyes confined quarters for bloodstock, it looks to have been light, convenient, and beautifully ventilated. Sharpe's racing colors, as listed in the records of the Philadelphia Jockey Club (of which he was a founding member), were yellow. They were often seen in the winner's enclosure in Pennsylvania and at Annapolis.

Governor Sharpe's was possibly our first polygonal stable, but similar schemes have since often been employed by amateur and professional American designers with considerable success. In later versions, of course, the straight stalls are replaced by loose-boxes; but box stalls, like inside plumbing, were rare in the eighteenth and early nineteenth centuries, and were generally reserved for rogue animals or those on the sick list. In our day, when horses commonly work shorter hours and are often fed energy-intensive rations, box stalls contribute to their health and safety, as well as help to stave off boredom and its attendant vices. Nevertheless, one cannot but suspect that the rise of the loose-box coincided with a rising standard of comfort among human beings and has, at bottom, as much an emotional as a practical rationale. Only the big draft breeds still commonly stand in tie stalls, and this seems neither to incommode them nor to ruffle their sublime serenity. Very few contemporary stables make allowances for anything but loose-boxes, yet numbers of straight stalls left over from an earlier day are still in use — and a great boon to the mucker-out they are.

43

IV Elegant Essentials

*In the afternoon, as we were driving rapidly along in the chaise, he said
to me, "Life has not many things better than this."*

— BOSWELL'S LIFE OF JOHNSON

By the latter part of the eighteenth century the eastern areas of the country
were served by a very respectable network of roads, many of which are still in
service. Under their modern dressing of asphalt and their aloof numerical desig-
nations, post roads and turnpikes still shoulder their burden of traffic as they
have done for two hundred years. The great rivers running to the Atlantic,
themselves busy highways, for some time largely confined the practical bounda-
ries of overland travel to within their watersheds; but within these limitations it
was possible to reach every civilized destination in a wheeled vehicle as well as
on horseback. There were mails and stagecoaches on the road, but public transportation was, on the whole, even scantier than it is today, and served only a
few trunk routes. No gentleman's residence, in town or country, could be complete without its domestic stable and carriage house, where the family's trans-
port was sheltered and fueled, and this was generally separate from any additional stabling for bloodstock or farm animals.

Some months before Horatio Sharpe returned to England, George Washington dined with him at Whitehall. He and his old friend doubtless visited the
Governor's racing stable, and Washington must have been struck by its advantages, for in 1792 he built a sixteen-sided brick and frame horse barn of his own
design at Dogue Run, one of the five farms of the Mount Vernon estate. His
drawing, and a very early photograph taken when the building was already
derelict, are all that remain of this interesting essay, which probably sheltered
mules, as well as horses intended for more utilitarian purposes than racing.

However plebeian, these animals were probably suited to their employment, for Washington was a knowledgeable and intrepid horseman, breaking
and training his own mounts to an exacting standard and taking a great interest
in breeding. He was responsible for bringing Lindsay's Arabian from Connecticut to Virginia; having seen some of the animal's get on a visit to New York, he

44

The stable at Woodlands (ca. 1790), Andrew Hamilton's "magnificent country place" in Philadelphia. Now a utility building for Woodlands Cemetery, it is described by Teitelman and Longstreth, in Architecture in Philadelphia, *as "perhaps even finer than the house, with its wonderfully simple use of Federal geometric design."*

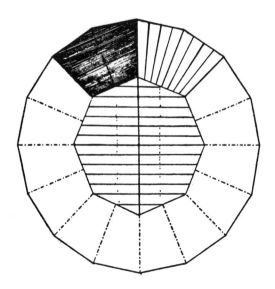

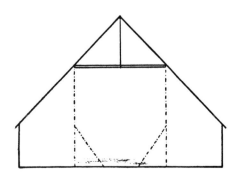

George Washington's drawings for the polygonal stable which stood on Dogue Run Farm, part of the Mount Vernon estate.

45

The 1782 domestic stable at Mount Vernon.

Interior of a stall chamber at Mount Vernon.

arranged the syndication of this mysterious survivor of shipwreck in his own state. The prepotent gray's blood is still traceable in modern pedigrees. Washington raced his homebreds with some success, and served as a steward of the Alexandria Jockey Club. When attending the races at Annapolis, he made it a principle to back Virginia-bred entries; a Maryland diarist has unkindly observed that this loyalty sometimes cost him dear.

The master of Mount Vernon also maintained a pack of hounds, established in 1767, and sometimes combined a foxchase with his daily tours of the estate farms. An entry in his diary tells of running a fox from eleven in the morning until almost three in the afternoon. Reynard's fate that day is not recorded.

Washington also designed the pleasant "split-level" brick stable which still stands at Mount Vernon. It was built in 1782 to replace a frame predecessor destroyed by fire. Although there is a separate coach house nearby, this building housed vehicles as well as horses, following the familiar convention of a central carriage-harness room flanked by two stall chambers, each in this case providing stands for ten horses. Out of sight at the back, on the lower level, there are eight additional open tie stalls, which constituted the mule shed for the home farm. Fodder was stored in the loft.

A 1785 visitor writes that the "venerable" Nelson and Old Blue, another of Washington's chargers, occupied adjoining stalls on the upper floor. "The General makes no use of them now . . . they feed away at their ease for past service." The same correspondent describes the stable as "nice," and rightly. Behind its quietly handsome facade it is tight but well ventilated (Washington insisted on more louvered dormers than his carpenter had apparently thought necessary), with a cheerful whitewashed interior and only a slight, but adequate, pitch to the stall floors. It would be a convenient barn in which to work, and must have made a comfortable retirement home for the two old campaigners.

As we have seen, the eighteenth-century gentleman was at pains to achieve architectural harmony in his surroundings, and was able to draw on a fund of available literature in its pursuit. The Miles Brewton House in Charleston, completed in 1767 by the London "civil architect" Ezra Waite, owes its great double portico and curved curtain walls to the Palladian influence. Waite and his employer, alive to all the latest English fashions, set off the classical lines of the residence with a charmingly Gothic carriage house, and thereby established a mode for Charleston designers. The front of this attractive little building is an exercise in the fantastical applied "Gothick" of designers Batty Langley, Thomas Lightoler, and Horace Walpole. The coach door opens onto the public street; the stables are an extension to the rear of the carriage house proper, making economical use of the limited available lot space.

Gothic outbuildings remained stylish in Charleston for many years, taking some interesting forms. The 1800 William Blacklock carriage house, for example, sports Gothic arches at door and window heads, which, in conjunction with the

A Frances Benjamin Johnston photograph of the Miles Brewton carriage house, Charleston.

The William Blacklock carriage house, Charleston, ca. 1800.

48

West elevation of the Governor William Aiken stable, Charleston.

double front door, give it much the air of a modest country church. In 1832, when South Carolina Governor William Aiken bought and remodeled the townhouse that bears his name, he also, it is believed, added the existing outbuildings, including the Gothic stable. Seventy years separate this building from the Miles Brewton carriage house, but subsidiary Gothic was still in vogue.

On the upper story, the stable presents seven blind Gothic windows to the street; the lower wall is pierced by arrow slits in the stuccoed brickwork, one to every stall. The twin louvered hay doors in the heavily corniced gable ends are also Gothic in form; the one to the south is rendered useless, except for providing a cross draft to the upper story, by its proximity to the small classical art gallery put up by Aiken to house that part of his extensive collections for which room could not be found in the house. Below the hay door at the north end is a massive, vertical-board postern hung on broad strap hinges. The wide carriage entrance and a narrow louvered grooms' door balance each other on the long east side. All three of these ground-floor openings carry a pointed arch, but the symmetry of the arrangement on the courtyard side was destroyed when a plain square-headed garage opening was cut in the east wall in this century; the square-headed windows on the courtyard side of the loft may not be original either.

49

Interior of the Aiken stable before restoration.

Within, the stalls are ranged along the street wall, with carriage and equipment rooms flanking them, under heavily boxed fascias supported on cylindrical pillars. The curve in each of the fascias comes to a gentle Gothic point. The wide hay drop, which runs the length of the stall range, is concealed by an inner fascia of the same design. Of the interior fittings, only the brick aisle, the somewhat mutilated upper structure of the stalls, and the skeleton of the low, doweled feed rack remain. Stall flooring, mangers, if they existed, and partitions have been removed. But one can still see enough to admire the excellence of the stable as it must once have been. The ventilation system, particularly important in a warm and extremely humid climate, is of special interest. The ceiling is high over stalls and aisle. There are generous exterior openings on three sides, and the second chimney (the first may have led from a stove in the carriage room) seems to have been intended as an exhaust flue from the loft, drawing off heated air as it rose through the hay drop opening over the horses' heads. The narrow slanted openings on the blind west side ensure a four-way circulation of air while preserving maximum insulation from the afternoon sun.

50

There is no cool place in Charleston in high summer (or was not, before the advent of air conditioning), but this stable must have been measurably more comfortable than the out-of-doors. Happily, the historic and noteworthy Aiken House and its dependencies have lately been acquired by the Charleston Museum, which plans to fully restore them.

If architectural fashions were tenacious in Charleston, they were slower to take hold and even slower to change in New England. Even Palladian classicism, which established itself so rapidly and so comfortably in Virginia and the Carolinas, was sparingly employed in the colder Puritan North, where occasional applied details like those on the Wadsworth stable long represented the extreme of classical character to be attempted in domestic architecture. Colonial New England carpenter-designers, or housewrights, generally relied on proportion and simplicity for their effects; one authority has stated that before 1776 no house in New England displayed so much as a four-column portico, let alone such a portico with columns soaring the full height of the house as at Whitehall or Mount Vernon. Outbuildings were, as a rule, built very plain indeed.

But the dawn of independence brought a new touch to the architecture of the northern coast. In 1774, the Port of Boston was closed, and there grew in Salem Harbor a fleet of the swiftest sailing ships in the world. They were first employed in illicit commerce and wartime privateering, and afterward in the most extensive foreign trade of any American port of the period. Before the War of 1812, fortunes were made in Salem. This wealth predictably found expression in architecture, much of which owes its superb quality to the genius of Samuel McIntire (1757–1811), a local housewright and wood carver.

McIntire came to his craft by inheritance, but his gifts for design and woodworking were of a rare order. He was employed in surrounding towns and even as far away as Nantucket, as well as in Salem itself; he has been called "The Architect of Salem" in tribute to the character he stamped upon that town. He was widely read in the available architectural literature, taught himself to draw with a facility unprecedented among early carpenter-designers, and applied his enlightened knowledge with confident skill. The earliest of his houses still standing was built for Jerathmeel Pierce in 1781, with details faithfully drawn from Batty Langley's *Builder's Treasury*. The courtyard stable owes its flavor to William Kent's scheme for the Royal Mews; the five-part Palladian facade is emphasized by pediments, heavy archivolts, and massive keystones, all executed in wood. The interior has suffered many changes over the years, but the arrangements appear to have been similar to those in the Wadsworth stable. Not surprisingly, the ventilation arrangements are adapted to cold weather; if necessary, all openings may be firmly closed, but the high carriage doors standing open under the lofty ceiling should have been sufficient to handle the occasional heat wave, with the assistance of the onshore breeze from the Atlantic.

51

The Jerathmeel Pierce stable and carriage house, Salem, Massachusetts, designed by McIntire in 1781; photograph by Samuel Chamberlain.

F. Maurer's 1749 engraving of the Royal Mews built for George II by William Kent, which once stood at Charing Cross.

In his later years, McIntire came under the influence of Charles Bulfinch of Boston. His outbuildings became less solid and weighty in appearance, but their basic design, having proved itself, remained unchanged; he lightened them with graceful examples of his woodcarver's art. The John Robinson stable, put up about 1825 by an unknown hand, received from its builder elegantly capped applied pilasters in McIntire's style at the corners. It was later further ornamented with carved rosettes and festoons taken from a McIntire coach house demolished in 1814, and by a pair of handsome rooftree urns taken from a church built by McIntire in 1804 and destroyed by fire in 1903. Unfortunately, the original interior plan of this building is now entirely a matter of conjecture.

52

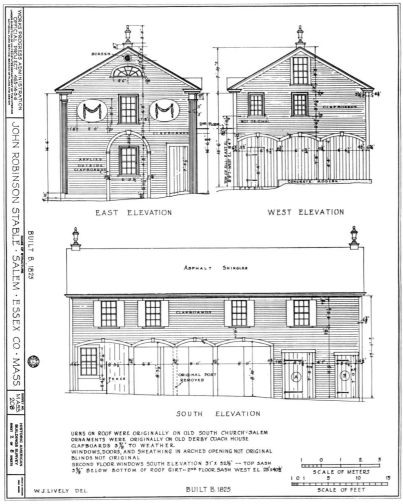

The 1825 John Robinson stable in Salem, which has been embellished over the years with applied details from lost McIntire buildings.

Barn and stable at Moor's End, a Jared Coffin house on Nantucket, in the style of McIntire.

EAST ELEVATION

WEST ELEVATION

ASPHALT SHINGLES

SOUTH ELEVATION

URNS ON ROOF WERE ORIGINALLY ON OLD SOUTH CHURCH-SALEM
ORNAMENTS WERE ORIGINALLY ON OLD DERBY COACH HOUSE
CLAPBOARDS 3⅞" TO WEATHER
WINDOWS, DOORS, AND SHEATHING IN ARCHED OPENING NOT ORIGINAL
BLINDS NOT ORIGINAL
SECOND FLOOR WINDOWS SOUTH ELEVATION 31"X 52½" — TOP SASH
3⅞" BELOW BOTTOM OF ROOF GIRT—2ND FLOOR SASH WEST EL. 25"X40½"

W.J. LIVELY DEL.

BUILT B. 1825

SCALE OF METERS

SCALE OF FEET

Left: Stable and carriage house at Gore Place, Waltham, Massachusetts, as remodeled by Charles Bulfinch in 1805.

Right: The carriage house and stable of the third Harrison Gray Otis house, Boston, designed by Bulfinch in 1807.

Charles Bulfinch (1763–1844) was our first native-born professional architect, although he was self-taught in the gentlemanly tradition of the aristocratic amateur of which Thomas Jefferson is perhaps our most illustrious exponent. Palladian classicism had barely touched New England buildings by the time of the Revolution, but it was certainly the dominant theoretical architectural fashion in America when Bulfinch toured in England and Europe in 1785. From his studies at Harvard and his own reading he was well grounded in *The Four Books of Architecture* and the principles of Vitruvius as interpreted by Palladio and Colen Campbell, yet the decorative and spatial possibilities of the romantic neoclassicism of Robert Adam and William Chambers, which he saw in England, gripped his imagination. He was sponsored by Jefferson in Paris, visited the Ambassador's favorite Paris buildings, and copied his European itinerary, but it was the concepts of the two English giants to which he remained largely faithful throughout his career. Between 1787 and his retirement in 1830 he achieved a varied and influential body of work. He designed the first playhouse in New England (the Boston Theatre); two penitentiaries; state houses for Connecticut, Massachusetts, and Rhode Island; the Boylston Hall and Market in Boston; the remodeled and enlarged Faneuil Hall of the nineteenth century; and a number of churches, banks, schools, office buildings, and courthouses. He created the first residential crescent in the United States, and a long list of distinguished private homes. It was Bulfinch who completed the United States Capitol, reconciling and improving upon the schemes of William Thornton and Benjamin Latrobe.

Few architects have left a comparable impression on our architectural land-

54

scape. Happily, among his surviving domestic buildings are at least one, and perhaps two, examples of equine housing. The commodious stable and carriage house at Gore Place, essentially pedestrian and utilitarian, probably received its impressive entrance detail from Bulfinch at the time he worked on the Gore mansion. The unchanged hay door and hook strike a slightly jarring note in this example of neoclassic facadism, and he very likely left the interior as he found it. But the stable and carriage house of the Harrison Gray Otis House at 45 Beacon Street, Boston, is a part of the original Bulfinch scheme.

This is the third Boston residence designed by Bulfinch for Mr. Otis, who was a successful and popular merchant, as noted for his princely mode of life and his lavish entertainments as for his considerable commercial sagacity. That the complex survives in such excellent condition is due to the efforts of the Beacon Hill Commission, under whose control twenty-two acres on the south slope of the hill are preserved from significant alteration; to the generosity of the Bulfinch Trust, founded and funded by the late Eleonora Randolph Sears, which transferred the property to the American Meteorological Society in 1958; and to the Society itself, which has displayed great sensitivity in adapting the place to its current use as Society headquarters. It is a sad little footnote to the project that Miss Sears, that gallant, versatile, and knowledgeable horsewoman, died only a few weeks before she was to have seen the completed stable renovation in 1960.

The building, which includes four stalls, feed room, equipment room and carriage area, closes off the courtyard at the rear of the house and is reached by a drive from Beacon Street. The court is bounded on the east by an annex wing that originally held the laundry, pantry, and shed, with serving kitchen and servants' bedrooms on the second floor. The third floor of the annex, which does not extend over the shed, is a later addition. It is now possible to enter the carriage house through the old shed, by means of a staircase that enters beside the stall range, using the space originally occupied by the privy.

The interior of the Otis stable today, showing the stall partitions and semicylindrical hay drops.

Stable and carriage house are now a projection room and exhibit space. The loft floor has been removed, exposing roofboards averaging over twenty inches in width; the mortise joints for the ceiling joists may still be seen in the cross-beams. The woodwork of the stalls, and their rounded corner hay drops, have been preserved, although their raked elm floors have been taken out and the cobblestones that lay beneath them reused in paving the courtyard. Mangers and racks are gone, but the tie rings, each set in its own granite block in the brickwork, are still there. Ventilation was provided through the hay drops and by means of a small exit into the lane opposite the grooms' door. But the stall dividers, rising almost to the level of the fascia across the front of the range, and the fascia itself, similar to if rather simpler than that in the Aiken stable, must have combined to shut out a lot of light from the stalls, even though keeping them snug and free of drafts in winter.

It was not easy for Bulfinch, in spite of his powerful and prolific talents, to support himself by his designs. Indeed, only his appointment as architect of the Capitol under President Monroe rescued him from perennial financial distress and allowed him to retire at last in reasonable prosperity. His creative ability was, not surprisingly, greater than his business sense, but he had more to contend with than a want of financial acumen. There was, in fact, too much competition, and not only from the traditional carpenter-housewright, who generally built to his own designs. It was quite usual in those days for a gentleman not only to design his own residence, but often to provide schemes for other private and civic structures. Much of our finest surviving architecture from before the middle of the nineteenth century bears witness to the gifts of these "dilettante" practitioners, Thomas Jefferson towering among them.

The variety and scope of Mr. Jefferson's talents have been exhaustively treated. His interest in horseflesh was doubtless subordinate to other concerns of this multifaceted genius, but it played a considerable part in his life nevertheless. Like Washington, he had an enviable reputation as a horseman in spite of his great height (in his youth a friend described him as a centaur). There is a story, possibly apocryphal, that a colt of his defeated George Washington's Magnolia at the Alexandria Jockey Club races in 1788. Be that as it may, Jefferson was not really interested in racehorses, although he did own a son of *Diomed, but rather directed his efforts at "improving the breed" to producing animals equally suitable under saddle or in harness and capable of light farm work. He owned and bred several animals of *Janus stock which proved highly successful; *Janus blood is traceable in Thoroughbred and Quarter Horse lines to this day. His horses were generally driven by postilions, one to each pair in the team, but if great speed were required he would take the reins himself. Many of his vehicles were built on the place, including gigs, phaetons, and a landau.

He had one stable at Monticello near the house of Edmund Bacon, his overseer, at the base of the mountain; another, put up in 1793, closed off the garden to the southeast of the mansion itself. This he described in his *Garden Book* as "105. feet long and 12. f. wide. one story high.," but the research now in progress on the street of shops and houses constituting Mulberry Row indicates that the building was L-shaped. In any case, the only stables yet restored at Monticello are the house stables, over the design of which he took considerable pains. He made an annotated sketch of Governor Penn's stable in Philadelphia, paying

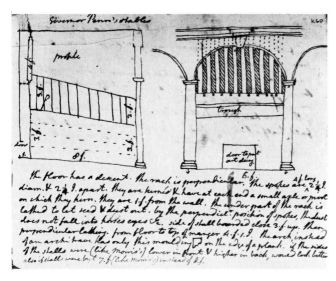

Thomas Jefferson's drawing (ca. 1778) of the interior of Governor William Penn's long-vanished stable in Philadelphia.

Detail from the 1772 Jefferson plan for the offices at Monticello, showing the stable layout.

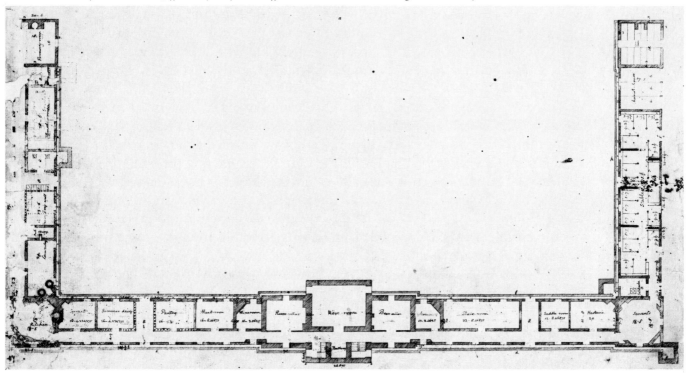

Mr. Jefferson's terrace stables at Monticello, Charlottesville, Virginia, built between 1778 and 1782.

particular attention to the ingenious construction of the feed racks, which he faithfully reproduced. The bottoms of these racks, slatted to allow dust to sift through, were behind the mangers, and their vertical dowels, set two and a half inches apart, rotated on pivots, so that reaching muzzles did not get caught. Jefferson took exception to the depth of Penn's stalls and the height of their partitions, which, he remarked, made the building dark, and corrected these defects in his own plans; his partitions drop off sharply to their low supporting posts. The stalls are ample in width — over five feet — but seem short, low, and sharply pitched to modern eyes. The floors slope to the aisle one inch in the foot, and the ceiling height over the mangers is just seven feet. This figure is not remarkable; high enough to allow a horse to carry his head normally, the ceiling is too low for the animal to injure himself severely by throwing his head or rearing. It is the length of the stalls, again barely seven feet, that is surprising, for many of Jefferson's personal horses stood sixteen hands and therefore had very little room to shift back and forth without stepping into the aisle. Yet the dimensions must have proved satisfactory, for the layout was twice reproduced elsewhere.

Jefferson was unable to install Penn's clean-out and ventilation traps under the mangers, as his stable has only one outside wall. Stall block, carriage rooms, and feed room are tucked into the depressed service wing under the north terrace. There is no through draft, but its cavelike situation and northern exposure protect the stable from extremes of cold or heat. Although a part of the house, and accessible from within it, the stable is totally concealed.

58

Pocahontas, *by Edward Troye, 1836. Behind Colonel Wade Hampton's famous race mare is his plantation house Millwood, near Columbia, South Carolina, with the up-to-date training stable to the right (the plan of which is shown here). Louvered ventilators ran across the front of the stall ranges, twelve feet up, to insure a draft-free circulation of air. There were sliding feed shutters in each stall, but in this version of a common labor-saver the mangers were in the aisle, and the horses put their heads out to eat.*

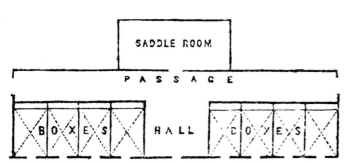

During his term as President of the United States, Jefferson designed and caused to be built the original White House terraces, accommodating the stables and other essential offices in the same fashion as he had done at Monticello. They were rebuilt, and a separate carriage house added, after the British burned Washington in 1814, but were left unfinished. Bulfinch, visiting the stables during Monroe's administration, describes them as "airy and well-ventilated," but notes "the ceiling of the colonnades is lathed but not plastered. . . . The appearance . . . is such that no gentleman of moderate property would permit it at his own residence." It is not known whether these wings ever did receive a finish before they were altered during Andrew Jackson's second term. Thereafter the White House stables were relocated, rebuilt, and remodeled a number of times. The 1857 stable was burned in 1864, destroying the Lincoln family horses in spite of the President's determined personal efforts to rescue them. When the blaze had died down, he was seen to weep for his dead son's pony as he looked out from the East Room at the smouldering ruins.

59

The last White House stable was converted into a garage during President Taft's administration and demolished in 1911. Ulysses S. Grant, breeder and driver of trotters, even "to the public danger," was the next to last true Presidential horse lover. Theodore Roosevelt, enthusiastic about horseflesh, history, and romance, backed Homer Davenport in the 1906 pioneering expedition to the Middle East which resulted in the first sizable importation of Arabian stock to this country, but Warren G. Harding rode in an automobile at his 1921 inauguration, an act which occasioned much bitterness. *Horse Review* animadverted sternly: "Every previous President of our country has either ridden on horseback in the inauguration parade, or else in a horse-drawn vehicle. . . . Senator Harding's break with tradition . . . is a curious incident in view of the fact that he was elected upon a platform which declared that our governmental methods had strayed away from the established traditions."

In the days when Jefferson "kept his horses in the White House," as one later report puts it, the American horse industry was in its glorious infancy, and bloodlines were established then that have come down to us today. The stables at Hampton, in Baltimore County, Maryland, are landmarks in this history (color plate 7). Horse racing and hunting played an important part in the social and sporting life of early Marylanders, even as they do now, and no family made a greater contribution to the equine annals of the state than the Ridgelys of Hampton.

"President's Levee, or all Creation going to the White House, Washington." The cartoon, from Andrew Jackson's first term, shows Jefferson's terraces as they were rebuilt after the War of 1812.

Photograph taken during the 1903 restoration of the White House by McKim, Mead and White, when Jefferson's wings were razed.

Hampton's fortunes were built on a foundation of iron. Colonel Charles Ridgely established the family ironworks. He was a founding member of the Maryland Jockey Club, and when he died in 1772 he left to his son considerable wealth, an estate that had grown to encompass seven thousand acres, and some fine blood horses. It was Charles Ridgely, Jr., who built the splendid Georgian mansion at Hampton, completing it in leisurely fashion between 1783 and 1790, an imposing evidence of the enormous profits derived from the family forges during the Revolution. Unhappily, this second Colonel Ridgely died childless within months of moving into his "Large New Building." Hampton and its riches passed to his nephew, General (and later Governor) Charles Curran Ridgely, who firmly established the primacy of the Hampton stud. Like his uncle, he was a keen hunting man, and was active in the affairs of the Maryland Jockey Club. He himself trained his racing homebreds. In his day, the ironworks supported a life of real splendor at Hampton, and his colors were made famous by such illustrious runners as Maid of the Oaks, Tuckahoe, and "The Great Maryland Horse" Postboy, grandson of Sir Archie, whose undefeated career spanned five years. In Washington, in 1798, Colonel John Tayloe III's Lamplighter bested Ridgely's Cincinnatus in a match race, but after the turn of the century Ridgely's horses were rarely defeated; they may earlier have contributed to Washington's losses at the Annapolis meetings.

The older of the twin Hampton stables dates from 1805. In his will, the Colonel had enjoined his nephew to build a stable "large enough to house six horses and six cows" for the dowager Mrs. Ridgely; it may have been built for this purpose by Charles Curran. In any event, it has been known since 1843 as "the racehorse stable." As the coach horses and their equipages were sheltered in two carriage houses elsewhere on the property, this southernmost stable must have been reserved for the valuable saddle and racing stock which was Hampton's pride; as early as 1800, an English visitor noted "Ridgely's horses, etc., of a superior sort and in much finer condition than many I saw in America."

61

Like the mansion, this building is solidly constructed of pink stuccoed rubble and stonework, but was later partly remodeled to match the 1847 stone stable, built to house trotting horses. At that time the layout was rearranged and four exterior doorways closed up. Save for the wood-floored corner tack room, the floor is of tamped earth over a crushed stone base. The walls are wainscoted to four feet, and finished in rough plaster above. The windows that opened onto stalls are furnished with iron bars diagonally set. The stalls — straight, of course — are generously sized, measuring nine feet, more or less, by five and a half feet. There is a central hay drop from the loft, below the cupola. Nowadays the building contains the crested 1850 Ridgely coach and a display of saddles that spans the two hundred years of Hampton history. The second stable matches the first, save that it contains only four stalls, roomy boxes with two windows apiece. By the 1850s the slow decline of the straight bay was beginning.

Governor Ridgely's wealth and influence gave him sufficient leverage in the racing world to swing the balance of power away from the old "racehorse country" of Virginia and North Carolina to Maryland, and within the state from Annapolis to Baltimore, where it remains to this day. His descendants, while never achieving the same uninterrupted string of racing triumphs, continued the family tradition of horsemanship. Hounds met regularly at Hampton until, in this century, the city came too close. The Maryland Hunt Cup was run there on four occasions, the first in 1895 and the last in 1920. Hampton became a National Historic Site in 1948, only then passing from Ridgely hands; it is now open as a House Museum. All the horses are gone, but an annual Pony Show is still held on the grounds, a happy contribution to a long and distinguished equine history.

The fortunes of Hampton were founded on iron, as precious and profitable a metal to the developing colonies as the gold and silver of Sutter's Creek and the Comstock were to a later generation. But generally speaking, the first half of the nineteenth century in the South was the era of the great agriculturists, and wealth was largely derived from rice, tobacco, corn, cotton, and cattle. The vast plantations, many of them standing on original Royal land grants, were very nearly self-sufficient communities, and often guided, as at Monticello, by forward-looking owners of cultivated intellect and interests.

General John Hartwell Cocke (1780–1866) and Thomas Jefferson were friends and (in the Virginia sense) neighbors, drawn together by mutual respect and a shared dedication to the values of scientific agriculture, general education, and architecture. When Cocke built his splendid Palladian residence at Bremo, overlooking the James River in Fluvanna County, he naturally consulted his distinguished colleague and incorporated several of Jefferson's architectural forms and devices. Cocke himself, however, provided many of the necessary plans and sketches, all of which were interpreted and built by John Nielson, a master builder who had worked on the remodeling of Monticello. In the out-

62

The stables at Bremo, Fluvanna County, Virginia, which serve the noted 1816 great house.

The Bremo Recess stables, which probably date from 1844.

buildings erected in support of his new home, Cocke experimented widely with forms and materials; the structures range from a mighty stone barn, carrying a noble Tuscan portico, to a small board-and-batten slave chapel with a modest Gothic entrance.

General Cocke was a man of fixed principles. He refused to grow tobacco, considering its use to be wasteful, dirty, and dangerous; he was also totally opposed to the use of alcohol. It follows that racing, with its attendant wagers, held no appeal for him, but he was extremely interested in his saddle and harness horses, and very particular as to their care. Like Jefferson, who purchased more than one animal from him, he was partial to bays. He had a favorite bay stallion called Roebuck, his charger in the War of 1812, and when visiting Jefferson, Cocke would put Roebuck in Edmund Bacon's charge because, said Bacon smugly, "he had rather trust him with me than with the servants."

The big Bremo stable is stone and brick, with a capacious loft. Arched hay doors in both gable ends are flanked by narrow louvered openings, suggesting tripartite Palladian windows. Louvered vents in all four walls provide cross drafts with the carriage and stable entrances on the north. The interior woodwork is black walnut and remains in an excellent state of preservation. There are two sets of rather small stalls with twin brick aisles at one end of the building; the other contains carriage and equipment space.

Left: The 1845 stone barn and stable at Belmead, Powhatan County, Virginia.
Right: The 1846 carriage house, barn, and stable at Roseland, Woodstock, Connecticut.

When the stables at Monticello were reconstructed in 1938, Jefferson's plan for them had not yet come to light, but Mr. Milton Grigg, the architect in charge of the work, discovered by measured drawing that each of the stall ranges at Bremo was a replica of those at Monticello, as indicated by archaeological research. The Bremo stalls thus served as a full-scale model for this portion of the restoration, an assumption triumphantly borne out by the subsequent discovery of Jefferson's own drawings and specifications.

The stone stable at Bremo Recess, where General Cocke lived while Upper Bremo was under construction, is in quite a different style. Built into a hillside, it presents, from above, a modest appearance, revealing little more than the loft space of the tall carriage blocks and the gabled roof of the long stable range, where the stall row opens onto a single aisle. In 1844 Cocke remodeled the original small frame house at the Recess in the Jacobean mode, with curvilinear brick gables, displaying curious Palladianized Gothic windows, and triple, diagonally set chimney stacks; but the stable building, impressive in scale and dimension as viewed from the lower access side, shows an austere lack of applied ornamentation.

In remodeling Bremo Recess in a Gothic mode, John Hartwell Cocke was in the vanguard of the romantic neo-Gothic movement in domestic design, which was born of the architectural eclecticism of the 1830s and continued to flourish until superseded by the didactic Italianate Gothic inspired (somewhat to his dismay) by John Ruskin, and now known to admirers and detractors alike as High Victorian. This midcentury Gothic has something of a fairy-tale quality, and was meant to charm the eye and catch the imagination rather than to exemplify an architectural philosophy. In 1845 General Cocke's son Philip commissioned the influential New York architect A. J. Davis to design his mansion at Belmead. It was the first Gothic plantation house to be built along the James, and was not universally admired; those who disapproved of the house were doubtless better pleased with the barn-stable, which is a smaller and somewhat naive version of the great Palladian stone barn at Bremo; the interior is much altered, but it probably had stalls for eight horses.

64

A few Gothic stables do survive from this period, and in their diversity we get a bird's-eye view of the flexibility of the Gothic mode as interpreted by the adventurous designers of the time. Roseland, in Woodstock, Connecticut, was built in 1846 to the plans of the fashionable New York architect Joseph C. Wells. A summer residence, it was named for its lovely gardens, but has always been better known as the Pink House, from the soft and cheerful color of its earthenware chimney stacks and vertical siding. It carries a restrained abundance of inventive Gothic detail: pointed gables, oriel windows, bargeboards and trellises, and other modish whimseys, all picked out in red. Now the property of the Society for the Preservation of New England Antiquities, it is in beautiful condition, as is the adjoining outbuilding, which combines the functions of carriage house, stable, and barn, and even includes among its amenities a bowling alley that is gently Gothicized within by the use of broad — but pointed — arches. House and subsidiary form a sophisticated essay in the Carpenter's Gothic taste, but Wells's work is thrown into the shade by the Wedding Cake House in Kennebunk, Maine, which is certainly the epitome of the style.

George Washington Bourne, who conceived and executed this breathtaking

The stall range at Roseland.

65

The 1852 ell at the Wedding Cake House, Kennebunk, Maine.

project, was a well-traveled master shipwright and yard owner. He admired "above all architecture" the Milan Cathedral. With careful draftsman's skill he drew its buttresses and spandrels, trefoils and quatrefoils, and in due course translated these carved stone details into his own xyloid vernacular. For twenty-five years Bourne had lived in apparent content in his foursquare Federal brick house with its modest barn and connecting ell, but in 1852 his retirement coincided with the loss of these outbuildings by fire. The replacements he erected received a fine dressing of gingerbread trim, all executed by Bourne using his shipwright's hand tools, and in 1855 he started to "fix the carved work on his house to match the barn," but on a far more ambitious scale. There is hardly an element of the medieval vocabulary that Bourne failed to include in his decorative schemes. The completed work is a masterpiece of Carpenter's Gothic, as much distinguished by craftsmanship as by vigor of design. The effect is heightened by the fact that beneath their lavish ornament the buildings remain stolidly unaltered, drawing nothing from Gothic building systems. The stalls, floor, and open mow of the stable are repeated on scores of New England farms, but above the high square-headed doors Bourne's astonishing slender pinnacles aspire toward some visionary architectural paradise.

66

Most of the Gothic flourishes so lavishly applied in the middle years of the last century were, like Bourne's, wholly extrinsic. But even the most romantic of styles can sometimes be employed to practical advantage. "A forted residence 'gainst the tooth of time/ And razure of oblivion," the stable building at the estate of Dunleith in Natchez, Mississippi, may be accurately described as a moated grange (color plate 11). The H-shaped pile rises out of a roomy paddock some little distance from the house. Gray masonry walls, topped by weighty crenelations, frowning double doors at the top of the carriage ramp, and heavy shutters at the few visible windows give the building a forbidding and inhospitable look, but the dirt-floored stable area, at ground level, is remarkably light, as well as cool, airy, and dry. Insulated by thick walls and the lofty upper story, it is ventilated by access doors and two additional windows at either end, and drained by the little ditch that hugs the foundation. Four masonry-built box stalls occupy the outer corners of the end blocks; open loose-boxes have replaced the stalls in the long aisle at a fairly recent date. Behind the castellated walls on the carriage floor there is ample room for vehicles, hay storage, and, very likely, at one time additional stall space. Whether the necessity for the drainage ditch dictated the Gothic flavor of the building, or whether a taste for the Gothic of the Waverley novels suggested the moat, the result beautifully combines the fanciful and the functional.

Sedgeley, Fairmount Park, Philadelphia, designed by Benjamin Latrobe for William Crammond in 1799. The house burned in the nineteenth century, but the stable (in the background of the picture) survives. It is used as a park utility building.

V *From Sea to Shining Sea*

The way to lift the mortgage is to hitch two good breeding mares to it and bid them go. — JOHN TUCKER

For the United States, the years between the achievement of independence from the British crown in 1783 and the formation of the Allied Forces against a common enemy in 1917 were consistently eventful. From a small group of intransigent colonies clinging to the eastern seaboard and often eyeing one another askance, the country developed into a vast, cohesive, and powerful nation, occupying the entire midsection of the continent. From the beginning, every new concept of technology was eagerly explored in order to forward the exploitation of apparently limitless natural resources. Progress was rapid and inexorable. Robert Fulton's sail-assisted steamboat first plied the Hudson River between New York and Albany in 1807, and by the mid-1830s side-wheelers appeared on most of the navigable tributaries of the Ohio and Mississippi. The first passenger railroad, opened in Baltimore in 1830, was horse-drawn, but steam-powered locomotives had long been hauling freight, and by midcentury passenger rail service was a commonplace. In 1869 track spanned the continent. In 1895 Charles Duryea patented the first successful American automobile, and in 1911 C. P. Rodgers completed a coast-to-coast flight in just over eighty-two hours. The harnessing of steam, and later of electricity, the development of the internal combustion engine, and the first practical applications of aerodynamics were the landmarks of this evolutionary process, yet through it all Americans continued to rely for basic, day-to-day motive power on the ox, the mule, and the horse — especially the horse.

Broadly speaking, the horses needed to perform the tasks set for them in the premotorized age were of three types. First was the road horse (trotter or pacer), suitable to ride or drive and capable of maintaining a smart pace with a modest burden. The Morgan, Saddle Horse, and Standardbred are the admirable results of breeding for these purposes. Second were the big draft animals, derived from heavy strains ranging from towering Shire to tidy Suffolk Punch,

68

One half of a stereograph view, ca. 1875, by Myron W. Pilley, showing the Grand Avenue horsecar barn of the Fairhaven and Westville Line, New Haven, Connecticut.

A street-railway car and a two-wheel dray at the entrance to a Philadelphia "manufactory"; a lithograph by W. H. Rease, ca. 1859.

69

An 1889 advertisement showing the Jordan Marsh & Company stable, where the haulage and delivery teams for the Boston department store were housed.

and often of mixed or highly uncertain pedigree. Slow, strong, and steady, they performed the weightiest tasks. Third was the speed horse — Thoroughbred, Arabian, or early "quarter-path" horse — which, with a light rider, could cover a long distance in a short time. This last classification includes the "remount": the cavalry and, later, police horse, breeders of which drew on Thoroughbred, Saddlebred, Standardbred, Morgan, Arabian, and even (at a pinch) mustang blood in the effort to produce a versatile, athletic, and tough-fibered animal of reasonably uniform and attractive conformation.

In the South and West, mules were used from the earliest days as draft animals; the size of the hitch varied with the size of the load. On at least one occasion in the 1880s, forty mules were put to one mechanical harvester. The humble burro is still unsurpassed as a nimble and enduring companion for travelers in arid wilderness areas; burros, mules, and ponies toiled underground and on slag-heap tramways in mines from Virginia to the Sierra (see color plate 10). But most of the equines that kept the wheels of commerce and communication turning for more than two hundred and fifty years were horses.

Horses hauled freight, pulled tramcars, drew stage and mail coaches, carried the horse soldier, the circuit rider, the eastern drover, and the western cowhand, and, briefly but gloriously, the Pony Express rider, on their appointed rounds. Now that the tractor-trailer, the bus, the van, the car, and even the motorcycle have almost entirely replaced these hardworking servants, their stables have fallen into disuse and to a large extent have disappeared. Only a few are still to be seen, preserved for their historical associations, converted to other uses, or simply awaiting demolition.

70

A building of this sort that has been particularly tenacious of life is the stable at the Black Horse Tavern, namesake of the vanished Philadelphia Black Horse Inn. It stands on the Bethlehem Pike in Flourtown, Pennsylvania. Its age is uncertain; the inn itself dates from the eighteenth century, but the stable is thought to be somewhat later. Probably the Tavern was an early stage or posting halt, but its greatest prosperity was derived from its situation as a convenient stopover for farmers who brought produce to Philadelphia markets, coming and going at regular intervals with their heavy wagons and multiple hitches. Later still the owners of the establishment had a coalyard, and the stable housed the haulage teams. It is many years since there was a stage service from Philadelphia to the West, Conestoga wagons do not ply the Pike, and even the coalyard is long a thing of the past. The stable, however, has successfully moved with the times, and now serves as a garage, with no further alteration than a minor amount of cosmetic repair and the cutting of an appropriate door into an aisle.

The interior configuration of the building is extremely interesting. At the end farthest removed from the road is a chamber containing seven straight stalls on the short wall and a generous aisle. These stalls are handsomely finished, with graceful dividers and turned posts. Next is a double stall range of very plain design, with a similar aisle (now the garage) between the rows. In every stall is a rather elaborate manger, consisting of a broad hay trough and a raised boxed shelf for a feed or water container. In the other half of the building is an open compartment with tie rings for fourteen horses, but no partitions or feed racks. A grain room lies beyond. There were no communicating doors on the ground floor, and no exterior door wide enough to admit a vehicle, although there is a wagon shed that runs out toward the road from the granary. Of the four hay doors, one has been closed up by siding applied early in this century.

Left: An 1880 etching of the innyard of the seventeenth-century Black Horse Inn, North Second Street, Philadelphia, of which no trace remains.

Right: The stable at the nineteenth-century Black Horse Tavern, Flourtown, Pennsylvania, from the road.

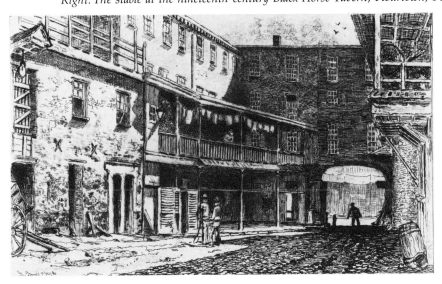

Traps and hay drops served the stall areas. There is a narrow hall stairway, probably of comparatively recent date, to the room above the granary.

Obviously the fee scale at the hostelry imposed certain class distinctions on the horses that lodged there. A hitching rack stood at the end of the Tavern, where local patrons could secure their horses. Rough-and-ready shelter was available in the tie room, and perfectly adequate if somewhat crude stalls in the next section; but the small end chamber offered a degree of refinement and exclusiveness in its select accommodations.

Later, the pattern of the livery barn or stage station, which gradually replaced the medieval-pattern innyard stable, came to show considerable consistency. As a rule, the stable floor was divided into two sections, with harness room and office (often combined), and perhaps feed room between them. Two wide doors opened onto the wagon shed on one side and a double row of straight stalls flanked a roomy aisle on the other. Stall partitions were heavy and solid, and the whole of the floor was generally planked. The loft was high and spacious, with a continuous hay drop over the mangers. Box stalls were emphatically not provided: the layout was calculated to handle the maximum number of transients expeditiously and with a reasonable degree of comfort. Of the three livery stables pictured here, the earliest is probably the one at San Juan Bautista. The configuration of the building, which is preserved as a museum, is practically the same as that of the one at Sandy Hook, but the irregular roof line has been concealed by the stepped false front. It is also the largest

Undated advertisement for a livery stable in Chester, Connecticut.

J. R. WOOD, LIVERYMAN,
CHESTER, CONN.
FIRST CLASS TEAMS
AT REASONABLE PRICES.
TELEPHONE CONNECTION.

The nineteenth-century livery stable of the erstwhile Sandy Hook hotel, Connecticut.

The 1874 livery stable at San Juan Bautista, San Benito County, California, rehabilitated and preserved as part of a state monument.

"Typical" nineteenth-century western livery stable at Old Tucson, Arizona. A popular tourist attraction, Old Tucson has provided the background for innumerable "horse operas."

surviving stable of the type that we have been able to discover. The hay drops here are of particular interest. They form a solid inner wall, sloping up from the mangers into the mow, and are broken only by pigeonholes at the bottom of the V, two to a stall, of an arched shape and size just right to accommodate an inquiring nose; it was easy enough to fill the troughs daily from the loft, and rely thereafter on gravity feed.

These years of rapid growth, with their inevitable attendant stresses, were frequently punctuated by bellicose incidents; the thirty-one years between the surrender at Yorktown and the tragicomedic War of 1812 was the longest period of unemployment the armed forces were to know, and between expeditions abroad and the almost ceaseless campaigns against the doomed Indian nations, the Civil War left its ineradicable stain. In these conflicts, the versatile, effective, and colorful horse cavalry played a major role.

Cavalry stables evolved in part from the livery plan. The death records of the Civil War are still, after two world-wide conflicts, difficult to comprehend. More Americans died in military service between 1861 and 1865 than in any military struggle before or since; only the heartbreaking trench warfare of 1914–1918 produced a higher percentage of fatal casualties, and the greater part of these occurred before the United States joined the Allies. The statistics on the loss of horses and mules on Union and Confederate sides are of course far from accurate, but the Northern armies lost an average of five hundred a day, and, while replacements could be had, the loss was scarcely less in the South. Equines were even more expendable than men, and almost as important in

Ranked stables at the mammoth Union cavalry depot at Giesboro, District of Columbia, in 1863.

The Riding Hall at Fort Riley, Kansas.

achieving military objectives, for not only the cavalry, but the artillery and the Quartermaster Corps depended on their services. Breeding stock of all kinds was decimated. Yet it was the War between the States that marked the beginning of the great days of the United States Cavalry, as it moved west to take and defend the center of the country from its original owners.

The central Union cavalry depot at Giesboro, in the District of Columbia, was the largest installation of its kind in United States history. It accommodated six thousand horses at a time, and was provided with a veterinary hospital, forges, storehouses, and full facilities for human personnel. In 1864 close to two hundred thousand horses passed through Giesboro. It was rapidly decommissioned after the war, and the acres of long stable buildings demolished. At this time, however, the pattern of the later cavalry barn became fixed, and although built of various materials, stables were thereafter almost identical in plan and elevation on every post.

They are long monitor barns, the slope of the roof extending out over twin aisles flanked with stalls, with one or two cross-aisles where equipment was kept and access to the open mow provided. Grain was kept in separate storehouses for daily issue; in some cases these granaries enclosed the area between barns to form a corral. It is a very effective and convenient plan and arranged for minimum wasted motion when the order came for "Boots and Saddles." Some of these buildings housed upward of a hundred horses apiece. In the long dry summers of the prairies the full-length vents under the monitors, the open aisles, the multiple high windows, and the wide doors at either end of the ranges must all have been needed to regulate the temperature. A handful of

75

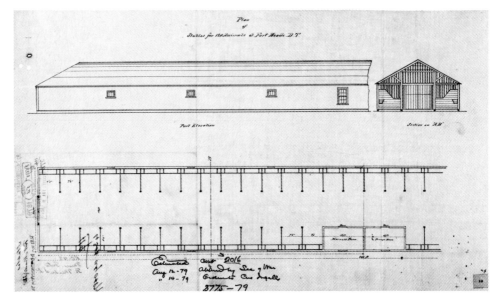

Working drawings, 1879, for one of the first stables at Fort Meade, South Dakota. The cost of building was estimated at $2,016.

The remaining stable buildings at Fort Meade in 1977.

these barns have been preserved with at least a part of the stable area intact (see color plate 8). The "troopers" that filled them — as many as a thousand on a big post — have mostly long since been forgotten, but the name of one legendary animal is commemorated in the stone stables at Fort Meade, South Dakota. Comanche's last rider, Captain Myles Keogh, perished with Custer in the 1876 Battle of the Little Big Horn, but this single cavalry survivor of the massacre came faithfully home by himself. Fort Meade survived as a cavalry post until it was turned over to the Veterans' Administration in 1944. It is now a hospital.

COLOR PLATE 8. *Nineteenth-century stables, Presidio, San Francisco.*

COLOR PLATE 10. *Stable for pit ponies or mules near the Comstock Lode mines, Nevada.*

COLOR PLATE 11. *Dunleith, Natchez, Mississippi.*

COLOR PLATE 9 (*opposite*). *Stable and carriage house, French Quarter, New Orleans.*

COLOR PLATE 13. *Domestic stable and carriage house, Belle Meade, Nashville, Tennessee.*

COLOR PLATE 14 *(opposite). Upper entrance to the five-level stable, Mohonk Mountain House, New Paltz, New York.*

COLOR PLATE 12 *(preceding pages). Leland Stanford's stick-style horse barn, Palo Alto, California.*

The Busch Place stable, St. Louis, Missouri.
Above: about 1890, before it was absorbed into the brewery complex. COLOR PLATE 15: *the interior as it is today.*

Travel by stagecoach also survived longer in the West than in the more populous East. The livery barn in Sandy Hook, Connecticut, may once have stabled relay teams, for it stood on a main route, but its chief business was probably with private travelers putting up at the adjacent hotel. The big barn outside of Lander, Wyoming, however, was built late in the nineteenth century to serve as a stage station. It is in a splendid state of preservation, complete with corral for spare horses. The towering loft with its penthouse roof, typical of local barn construction, is a reminder of the isolation that can still seize ranches and farms when the snow comes in from the north.

As urban centers grew, city stabling became a vexing problem. Cab horses, fire horses, ambulance horses, tram horses, milk and bakery horses, and the teams that drew the drays and wagons for brewers, millers, freight haulers, and express companies, shared the crowded streets with private carriages and the myriad equipages that delivered retail merchandise. In smaller communities, stabling was simply attached to carbarns or commercial establishments as needed, but in the cities, with land at a premium, more inventive shifts were required. The high-rise parking garages of the present day had their prototypes in multistoried stables, with ramps between the floors. A few of these ingenious equine tenements are still in use. Although the system seems harsh and even dangerous to the unaccustomed eye, many hard-working animals have lived to a great age in these establishments, apparently thriving on hard and slippery pavements, long hours of labor, smoggy air, oppressive heat indoors and out in summer, and climbing upstairs to bed at the end of the day (see color plate 9).

Other city stables were built partly below ground, like those of the New York City Police Department in Central Park, which are still very much inhabited. Others occupied the space in the center of urban blocks. A few, like the octagonal barn at the Anheuser-Busch Brewery in St. Louis, were built as private stables and turned over for commercial use when the city engulfed the

The stage station at Lander, Wyoming.

The Pony Express stable at St. Joseph, Missouri, now part of a museum complex. The monument in the foreground marks the point of departure of the first Pony Express rider.

The vanished Hibernia Fire Engine Company No. 1, York Street, Philadelphia, 1852.

The stable of the Polk Sanitary Milk Company, Indianapolis. Delivery wagons occupied the ground floor; the stable, reached by a ramp, was above.

original house. The Busch stable survives as a showpiece, and is used only when the famous Clydesdales are in residence. It is a handsome period building, and beautifully fitted inside, but a great part of its charm lies in its location in the lowering shadow of the tall brewery buildings, with a network of vast conduits crisscrossing in mysterious patterns overhead.

Firehouses and hospitals made particularly ingenious arrangements in stabling their teams for the rapidest possible mobilization. The stalls were so situated that the horses could be backed directly into position in front of the vehi-

Now derelict, this Mobile, Alabama, firehouse probably held both hand- and horse-drawn equipment.

Engine Company 23, San Francisco (1893), last survivor of the city's Victorian firehouses and now a residence.

An 1886 photograph of an ambulance in front of its quarters at Manhattan Hospital, New York City.

The 1913 horse barns at the Springfield, Illinois, State Fairgrounds.

cle, and the harness was suspended overhead so that it could be dropped into place and fastened with maximum dispatch. Bridles were hung so that they could be put on before the animal was backed from the stall; the fire teams or ambulance horses could be made ready almost before the men could board the engine or the big doors be flung open.

Stable arrangements at state fairgrounds reached their apogee after the turn of this century. The raceways which still play so large a part in fairground activities have of course always had their own stabling, built either around an exercise area for winter workouts, or in the more familiar shedrow style, but fire or the risk of it has swept away most of the older versions of these. In the central portion of the fairgrounds, however, where prize livestock of all kinds are judged, splendid show barns were raised for quite a different kind of horse.

After the Civil War, there was an active and increasing interest in purebred draft stock. Imported Shire, Clydesdale, Belgian, and Percheron sires were bred to local and imported mares with impressive results. Breed classes at the local and state level were large, and championships well earned. These animals were intended for work as well as show, and a premium from a knowledgeable judge may have had a telling effect on stud fee or sales value. The trotters, deservedly glamorous, drew the biggest crowds, but the in-hand classes were a thrilling sight, and money in the bank to the owner whose huge horses, gleaming from the wisp and sporting knots of ribbon in the old style, left the arena as champions. No straight stalls for these potential superstars, at least at the fair, but large and handsome box stalls, well-bedded in golden straw, in buildings of the most approved architectural style.

As horses came to play a smaller part in the agricultural process, these classes dwindled, and the gracious stalls came to be occupied by lighter breeds entered in the performance classes of more modern horse shows. Many of the

80

The enormous Coliseum stable at the New York State Fairgrounds in Syracuse. The later stable annex (not shown) is very nearly as large.

oversize stalls were fitted with tailboards to protect the high-set flaunting banners of Three-Gaited, Five-Gaited, and Fine Harness American Saddlebreds, which were beginning to carry their smart appearance, confiding manners, and comfortable way of going from highway and plantation to the show ring.

The big breeds by no means disappeared, but it would be fair to say that by midcentury comparatively few of the younger generation, save for those in truly rural areas, could claim to have seen any purebred heavy harness horses, aside from bareback riders' mounts or perhaps the Clydesdales of the touring Budweiser hitch. Now the number of these kindly giants is again on the rise. Belgians, Clydesdales, and Percherons remained well established here, but Shires and Suffolk Punches are being brought back, with the addition of such less well-known breeds as the Jutlander and the Netherlands Draft Horse. The reason for this revival may in part be sought in our new ecological awareness. It is true that the big horses eat a lot, but they also do a great deal of work with a minimum of noise, can work a field when a tractor will bog down, and make a contribution, both in labor and fertilizer, to the production of their own and their owners' food supply. Finally, unlike more sophisticated machinery, they can reproduce themselves, thereby constituting a renewable resource that can also turn a nice profit — especially after winning a blue ribbon at a fair.

81

VI *No Expense Spared*

A horse should be treated like a gentleman.
— LELAND STANFORD

American history between the surrender at Appomattox and the sinking of the *Lusitania* has been variously subdivided at the convenience of historians. Catch phrases like Reconstruction, Age of Steam, Western Expansion, Gilded Age, Victorian (or Edwardian) Era carry rather specific connotations in terms of political, social, economic, and architectural history, each describing some aspect of what was, in fact, the American Age of Affluence.

The legends of the enormous fortunes amassed during those years have been recounted perhaps too often; the sagas of people exploited, along with dazzling natural resources and developing technology, by the founders of some of these fortunes are also unhappily familiar. But what is still difficult for us to grasp is the enormous buying power of that wealth. Even a humble monomillionaire was possessed of untaxed funds that challenged dissipation. At the same time, labor was cheap, food was cheaper, and goods and services of all kinds were to be had for sums that seem to us unbelievably modest, even when calculated in gold-based currency. To create some kind of dynastic monument is an especially satisfying form of conspicuous consumption; instant ancestral mansions became the order of the day. All but a handful of our most expansive, elaborate, and superbly appointed stable buildings were created even as the iron horse that helped to pay for them was putting its mortal cousins out of employment.

Adolf Weske was a Californian of Russian descent, a successful entrepreneur who owed his prosperity to the boom-and-bust economy that followed the gold strike of 1849. His San Francisco residence was a suite at the Palace Hotel, but he also owned a country retreat outside Healdsburg, in Sonoma County. In 1868 he put up a stable on the property to house his trotting-horse mares and foals and his foundation stud, Black Prince. It is an outstanding building, much resembling in design and construction the so-called Russian barn at Fountaingrove in Santa Rosa, Sonoma County, except that it has two high wagon doors

82

The 1868 octagonal stable at Mount Weske, Sonoma County, California.

The quondam Fountaingrove Community barn, Sonoma County, which was built according to the same structural systems as the Mount Weske stable.

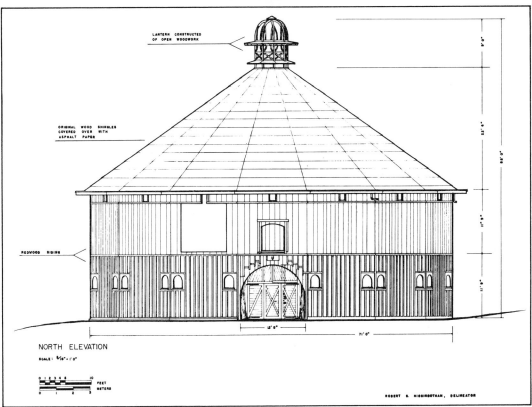

LANTERN CONSTRUCTED OF OPEN WOODWORK

ORIGINAL WOOD SHINGLES COVERED OVER WITH ASPHALT PAPER

REDWOOD SIDING

NORTH ELEVATION

SCALE: ³⁄₁₆" = 1'-0"

ROBERT G. HIGGINBOTHAM, DELINEATOR

rather than one arched opening. Octagonal in plan, it measures one hundred and five feet across and one hundred and five feet from the central fountain and drinking trough to the roof peak. Four blocks of box stalls, with a closed aisle behind them, face inward. They are separated by the door aisle, and by office and living quarters on one side and two stallion or foaling boxes on the other. The entire building is constructed of redwood, save for the mangers, hayracks, and finish rails in the stalls, which are of European cedar (the delectable quality of redwood to the equine palate was already obviously well known). The floors are dirt, graded and raked.

83

There was no telegraph to Mount Weske in the builder's day, although the railroad ran to Healdsburg. When Mr. Weske planned to visit he released a pigeon, message attached, from the Palace Hotel, and it flew to the pigeon loft in the cupola, ringing a bell by its arrival. In order to find out which train to meet, the manager had to scale the open stairway, lightly railed on one side, that rises in a series of dizzying sweeps from the open loft over the stalls to the peak. Like the Fountaingrove barn, this stable is said to have been built by Russian sailors; perhaps the stud groom had spent years ascending to the foretop before settling down on land. No other buildings survive at Mount Weske, nor did the original owner leave much trace on history, but his stable still stands. Virtually unaltered, but refurbished and painted in its original colors, it is now the headquarters of Mount Weske Arabians, which is under the direction of Roy and Inga Applequist.

A founding partner in the Central Pacific Railroad, Governor (later Senator) Leland Stanford (1824–1893) reaped a golden reward for his foresight and industry when the tracks were joined between coast and coast. In 1875 he acquired a huge ranch on the San Francisco peninsula at what is now Palo Alto, and promptly entered with enthusiasm into a large-scale stockbreeding and training operation. He raised Thoroughbreds, joined with Lucky Baldwin (see page 86) to found the Pacific Bloodhorse Association, and built a racetrack, the Bay District course, near San Francisco; but his real love was trotting horses.

In 1876 he purchased thirteen trotters from Charles Backman of the famous Stony Ford Farm, in Goshen, New York. Among them was Electioneer, an unraced son of Hambletonian. Electioneer sired one hundred and sixty Standardbred performers in his fourteen years at Palo Alto. (In 1879 the three-year-old National Association of Trotting Horse Breeders established the first criteria for the registry, setting a minimum standard of 2:30, 2:35, or 2:40 for a trotted or paced mile, according to age and sex, and whether the record was set under saddle or "to wagon." The standard was rigorously applied to registrants or their immediate progenitors. Hence the "Standardbred.") Two of Electioneer's get trotted the mile in less than 2:10.

Leland Stanford had an inventive and inquiring mind, traveled widely in Europe with his wife and only child, and kept abreast of contemporary developments in science and architecture. He did not race his own horses, although he drove trotters for pleasure; his interest was in improving bloodlines, performance, and training methods. He was successful in upgrading his stock by introducing judiciously selected Thoroughbred mares into his pedigrees, and promulgated a revolutionary training system, the basic tenets of which — start as a yearling, train fast, and win races in as few heats as possible — are still applied.

The Stanfords' son, then almost of college age, died of typhoid in Europe in 1884. His parents determined to establish a university in his name on the ranch that would never have an heir. Leland Stanford Junior University still owns eight thousand semipastoral acres in the midst of suburban pressure and still strives to preserve the Jeffersonian "academic village" principle settled upon by Stanford and his chief adviser, Frederick Law Olmsted. The central ranch buildings were almost all swallowed up in the project (one stable survived for a time as a student cafeteria), but there are still two stables intact. Designed in unrelated styles, but with equal attention to architectural and practical values, they are a reminder of the great days of the Palo Alto Stud.

One of these, now vacant, is a pleasant stuccoed building once used for polo ponies. The other, a high, cupolaed, red and white, stick-style barn that dates from the founding of the ranch, was a training stable for trotters (color plate 12). It is striking for the elegance of proportion and detail which lend it a distinction unusual in such a large and basically mundane building. Inside, the wide aisle and big stalls have been somewhat altered to cope with new needs, for the building now serves the University riding operation, but it still shows the spacious and workmanlike pattern common to Standardbred barns of the period. It was on the small track adjoining this stable that Eadweard Muybridge, with Stanford's backing and cooperation, took the photographs for his historic work *The Horse in Motion*, which revolutionized common conceptions of how a horse moves at speed. This study no longer astonishes eyes accustomed to split-second exposures and slow-motion cinema, but its production in that day was little short of a miracle, and its effect on graphic art and systems of horse training have extended far beyond those who have seen the book or are even aware of its existence. In 1887 Muybridge published the encyclopedic *Animal Locomotion*, lately reissued, in which his equine studies are included.

Illustration from Eadweard Muybridge's 1878 Animal Locomotion.

Another California tycoon of the period was Elias Jackson Baldwin (1828–1909). He was known as "Lucky" Baldwin, but much disliked the nickname. True, the Comstock Lode brought him a fortune, but he relied for success on hard work and sound planning; and indeed it was the land into which he poured his silver that repaid him in the end. Even his racehorses consistently showed a substantial profit. He paid two hundred thousand dollars for eight thousand acres of Rancho Santa Anita, unfashionably located in southern California, in 1875, and in time expanded the property to more than seventy-five thousand acres.

Baldwin's homebreds raced — and won — all over the country. One of them, after a successful career in the States, raced in England for Richard "Boss" Croker under the name Americus. Americus sired the English mare Americus Girl, who founded an impressive female line in England that was instrumental in the final acceptance of American Thoroughbreds in the General Stud Book. Three others won the American Derby in Chicago, and Emperor of Norfolk, purchased for his Lexington blood, won races in five states and the District of Columbia. In his eightieth year Baldwin founded the first Santa Anita Track. It closed after two seasons, when antibetting legislation crippled California racing, but six American records were established there in 1908 and 1909.

More than a stud farm, Santa Anita was a showplace, and more than either it was a self-sufficient and commercially productive complex. Baldwin spent lavishly on beautification (it was Mark Twain who remarked, "The best is none too good for Baldwin!"), but far more lavishly on practical innovations and improvements. He raised all the hay and feed for his horses, cattle, hogs, and poultry. He established orchards, nut-tree plantations, truck gardens, vineyards, and a winery and installed an extensive irrigation system to serve them. He built and operated several hotels, including two grand resort hostelries, all of which were wholly provisioned from the ranch; the balance of the produce was sold at a profit. When the Comstock failed and the Bank of California went down with it, his solid investment in Rancho Santa Anita kept Baldwin triumphantly afloat.

The owner's residence at Santa Anita was the 1839 adobe ranch house put up by Hugo Reid when he patented the original property. When Baldwin first stayed there, his carriage horses must have been taken off to the racing stables (long since vanished) or sheltered under simple tule trellises. But by 1881 his ornate gingerbread guesthouse, endearingly miscalled the Queen Anne Cottage, was complete, and adjoining it he built the Coach Barn, similar in exterior style to the Cottage and beautifully finished inside in alternate redwood and cedar vertical boards. The four generous tie stalls have a large window apiece, and share semicylindrical cedar and redwood drops above handsome iron hay racks. The floors are of wood, with an ornamental iron drain cover dividing the stalls from the aisle. The partitions, capped with graceful grillwork, are set into sturdy posts with turned finials, and the cedar mangers are rubbed smooth. The

*Lucky Baldwin's carriage house,
Santa Anita, California,
complete with Gothic doghouse.*

*The stall area in the
Baldwin carriage house.*

building held the tally-ho and the matched team that drew Baldwin and his guests from the railroad. Hearsay has it that these were imported English Hackneys; if so, they can have felt no diminution in comfort and style in their New World quarters. Even the stable dog (can he have been a Dalmatian?) had his own Carpenter's Gothic doghouse under the eucalyptus by the carriage door.

Rancho Santa Anita, of which one hundred and twenty-seven acres remain, is now operated as the Los Angeles State and County Arboretum. The plantings, carriage house, Queen Anne Cottage, and Hugo Reid Adobe have been preserved and restored under this sympathetic aegis; all are open to the public.

Adolf Weske and Leland Stanford were trotting-horse men; Lucky Baldwin loved fast-running Thoroughbreds; but elsewhere in the United States were those dedicated to the preservation of the unique equine line descended from Justin Morgan, and the most dedicated of them all was Colonel Joseph Battell. Battell compiled the original Morgan Horse Register, ruthlessly excluding those animals in whose pedigrees too many outcrosses appeared from the founding strain, and in 1907 gave the United States Department of Agriculture a four-hundred-acre tract in Vermont, together with the magnificent 1878 barn he had built there, on which to operate a horse farm where the true Morgan would be perpetuated and preserved. He also presented the establishment with its herd sire, General Gates, a stallion of his own breeding. General Gates, like his full brother Lord Clinton, the fastest Morgan trotter ever developed, was of impeccable lineage, combining the Justin Morgan bloodline with that of the Thoroughbred *Glencoe, who appears in the ancestry of an astonishing number of horses of all American breeds. The experiment was highly successful, and although the farm was transferred to the University of Vermont in 1951, since the Department of Agriculture no longer had any responsibility to maintain remount and police-horse types, the university has continued to protect the integrity of this remarkable breed while conducting valuable equine research on nutrition, productivity, and development. There is hardly a Morgan breeder who does not have strains of the great sires and dams of the Experiment Station in his herd.

The big Weybridge stable is still in active service. It is a bank barn, and the marble foundations supporting it were quarried at nearby Proctor. The roof is Vermont slate, that of the mansard still the original material. The loft is breathtaking in structure and proportion, from the great braced timbers supporting the massive weight of the roof to the heavy and close-set joists and rafters that make it possible to store a year's supply of hay for seventy horses, all that the farm currently supports, in this single space. The cupola, reached by a ladder stair, is crowned with a gilded cast-iron weather vane, of which only eleven were made. On the entrance floor, with its high ceiling, there are seven box stalls, floored in triple elm planking, laid green, and jacked together for a perfect fit. The center is a large carriage room, now a reception center for the thousands of tourists who visit each year. The original washroom has been converted for tack and harness storage, and the old tack room turned into a projection room. On the lower floor there are seventeen box stalls opening onto a generous aisle. All of these are floored in clay, although three have a cement base. The ceiling downstairs is not so high as the one on the upper level, but still measures between nine and ten feet.

Morgans are bred all over the country today, for show or simply for pleasure; they are still in demand for mounted police work, as well. Thanks to Colonel Battell and the Experiment Station at Weybridge, the plucky, muscular animal immortalized in the statue of Justin Morgan that stands in front of the Weybridge barn still gives pleasure to horsepeople.

The grand barn at Weybridge, center of the University of Vermont's Morgan Horse Farm.

Main breeding barn at Shelburne Farms Stud, near Burlington, Vermont, where Mr. Seward Webb raised English Hackneys, French Coach Horses, and Hackney Ponies.

Lucky Baldwin was not the first to build grand resort hotels. During the 1870s and 1880s, gigantic and luxurious summer lodgings proliferated, rising with seemly dignity, but in astonishing numbers. They were generally served by rail, which brought in clientele and supplies alike. Many were associated with mineral springs or spas, but others relied simply on good food, comfortable accommodation, attentive service, and scenic beauties in refreshing air to attract their visitors. Providing a comfortable holiday residence, catering to real or imagined physical difficulties, and supplying endless opportunities for polite social intercourse and well-organized matchmaking, they flourished. Comparatively few remain; their ranks have been decimated by the ravages of time, fire, altered transportation patterns, increased labor costs and land values, and the general decline in public formality.

Two that have triumphantly defied the penalties of anachronism are The Homestead, in Hot Springs, Virginia, and the Mohonk Mountain House, in New Paltz, New York. They are very different from each other, but have similarities which may hold the secret of their survival. Both were founded by families who have maintained an active interest in their operation. Each has successfully kept up with the changing requirements of a contemporary year-round clientele, has taken vigorous action to protect the quality of the environment, and has been at pains to preserve the idiosyncratic charm that attracts the newcomer and brings back old friends.

Among the amenities generally included at the better class of resort establishment were stabling facilities, providing saddle and carriage animals for hire as well as sheltering visitors' horses. Two outstanding examples of this special

90

*The Criser stable,
Hot Springs, Virginia.*

form of livery stable are still in operation at Hot Springs and Mohonk. At both hotels the visitor has access to miles of wood roads, winding through the soft Virginia mountains or the precipitous Shawangunk formations above the Hudson. Automobiles are forbidden, but walkers, trailriders, and horse-drawn vehicles share the paths without prejudice.

The Homestead stands on the site of a modest spa facility founded in 1766 by two veterans of George Washington's French and Indian War campaign, and long since lost in the foundations of the present gracious Victorian pile. Its stable is down in the village of Hot Springs, only a little distance from the hotel by bridle path, but well removed by road. The spacious frame buildings, which, with their modest stableyard, fill one of the few level plots in the area, are laid out in a double H. This allows for cross-ventilation to every double range of stalls, and permits each section to be put into use or withdrawn as required. The stables are unoccupied in winter, and are therefore very airy, high-ceilinged, and dirt-floored; the stall dividers are slatted. Rides or drives start from the hotel, so there are no trimmings and no upholstered waiting room; but everything necessary to run a busy commercial stable is present and in good order. The building and the horses are owned by the Criser family, the third generation of which now supplies mounts and carriage rides to Homestead patrons, under contract with the hotel. Horses are rarely purchased for this operation; most of them, of Thoroughbred or Quarter Horse stock, with a few Tennessee Walkers and draft crosses, are bred, raised, and schooled by the Crisers to ensure the desired qualities of soundness and temperament.

Mohonk, founded in 1870, is gloriously eclectic in exterior architecture and

91

The architect's drawing of the west elevation of Belcourt, Newport, Rhode Island. The stable wing is to the right.

dedicated to simple comfort within (color plate 14). The stable belongs to the hotel. It has saddle horses and horse-drawn vehicles (with drivers) for hire; it buys or leases its animals. Flat land is also at a premium here, and the five-story stable is stacked into the hillside beyond and below the hotel, with access at ground level on every floor but the loft. At the top, facing west, are the carriage house and highest stall ranges, which contain box stalls; the loft is above. On the floor below, where access is to the north, three aisles of straight stalls surround a grassy quadrangle. The second floor opens to the south, and on the lowest level the doors are on the east. There is an interior stair between floors. All floors are oak or elm, sharply raked in the wide, high-partitioned straight stalls, where the boards are laid perpendicular to those in the generous aisles. Difficult to describe, and impossible to sum up in a photograph, this is an adventurous and successful building. It is staunchly built and designed to handle a busy service operation; it has proved efficient and comfortable even when, on the occasion of the American Driving Society annual meetings, a score or more of equipages are "putting to" there at the same time.

One of the earlier of these tremendous summer hotels, irreverently described in its heyday as a "huge, yellow, pagoda factory," was the Ocean House at Newport, Rhode Island. It was also one of the first to fade, and burned down in a state of neglect in 1898. Long before that year, Newport had been transformed from an unpretentious retreat for the genteel artist and intellectual into the golden enclave of monied society. Such persons did not live in hotels, but in residences; the more overwhelming the residence, the better.

92

One of the most remarkable of these marvelous "cottages" is Belcourt, erected in 1892 for Oliver H. P. Belmont by the brilliant, eclectic architect Richard Morris Hunt. Belmont's father, Augustus, who came to this country in 1837 as a representative of the Rothschild interests, became a social leader. President of the New York Jockey Club, he founded the Nursery Stud on Long Island, where he raised Standardbreds, largely of the Hambletonian line, and Thoroughbreds of imported and American stock, and kept successful racing stables. Augustus II, buying in much of his father's foundation stock from the estate, re-established the Nursery Stud in Kentucky; in 1918, Samuel D. Riddle paid five thousand dollars for Belmont's homebred yearling Man o' War. August II's

Aerial view of Belcourt.

93

brothers shared the family interest in horses and, to a degree, in racing, but both especially enjoyed participating in their horse sports. Perry was a talented equestrian; Oliver cared most for his driving horses. At Belcourt, loosely patterned after a Louis XIII hunting lodge, stables and coach room are part of the house. Stylistically, the mansion is a fascinating agglomeration, over which a brooding Gothicism seems to lie. The medieval feeling was intensified by the formal carriage entrance, which allowed vehicles to pass through the house proper.

In 1896 the bachelor Oliver married the former Mrs. William K. Vanderbilt. Like Hunt, Mrs. Belmont had studied at the École des Beaux-Arts, one of the first women to do so. After her husband's death in 1908 she made a number of architectural and decorative changes at Belcourt, in the course of which the interior carriage drive was blocked off. The stable itself is not much altered; twenty-six tie stalls, partitioned in teak and with modish grillwork, and an additional loose-box at either end were set against a tile wall picked out in the Belmont racing colors of maroon and scarlet. At the farther end is an ample carriage room, with grooms' quarters above the whole. When the stable was occupied, each stall had its plaited straw border, every horse wore a fresh white linen cooler displaying the family crest, and the handmade English bricks in the aisle were covered with a layer of clean sand, stenciled with sketches of coaching scenes that ran the length of the one hundred and sixty-foot aisle.

Belcourt is presently owned by the Tinney family, who have preserved and restored the house as a setting for their extensive collections. The house is open to the public from May to November, but the stable area, currently given over to storage and to Mr. Tinney's work in stained glass, is not on view. The owners hope, however, eventually to share its refurbished elegance with visitors.

Including stable and house under one roof was not in itself unusual at the time. In 1886 H. H. Richardson (1838–1886) built a Chicago residence for Mr. and Mrs. John J. Glessner which, as the last and finest of that seminal architect's residential works, is a landmark. Presenting a forbidding facade to the street, it embraces an open courtyard to the rear, flanked on one side by the kitchen wing and on the other by the stable block. The slit-shaped windows, the arched center entry, the high, square-headed entrance to the coach room, with its door recessed the depth of the massive wall, the broad hay door and beam above it, and the cote-like cupola topping the mow combine to lend the entrance front the air of an earlier and more defensive time; but on the garden side the house becomes open and welcoming, the walled garden free of intrusive structures. The stable proper contains four straight stalls, two loose-boxes, and a tack and harness room, with a washrack in both carriage room and stable aisle, and a window to every stall. Like the rest of the house, the stable is an elegant and efficient accommodation which reveals nothing of its warm character to the outside world. The house is now in the hands of the Chicago School of Architecture Foundation.

The Glessner stable from within the courtyard.

Contemporary sketch of the elevation of the Glessner carriage house and stable, Chicago.

At Frederick Law Olmsted's suggestion, Richard Hunt incorporated the stables into the fabric of Biltmore, George W. Vanderbilt's enormous château in the mountains of North Carolina, which was started in 1890 and completed after the architect's death in 1895. At Biltmore the stables (now a workshop) form part of a courtyard composition which also served the dairy and machine shop, consolidating the estate operation and incidentally enlarging the vast building. Yet despite their proximity to other domestic quarters, the Glessner and Biltmore stables are distinct entities. At Belcourt the horses shared the residence of their affectionate bachelor owner on almost equal terms.

95

The stable at Biltmore, North Carolina, from above. This part of the estate will be opened to the public in the spring of 1982.

Shown here and opposite are two Gilded Age stables in Texas.
This stable and coach house was designed for Daniel Sullivan by the English-born architect Alfred Giles in San Antonio. The building is limestone, finished in hard pine within. It contains eight stalls, carriage room, harness room, cleaning room, and loft living quarters, with feed storage in the mow. Now owned by the Hearst Corporation, this rather splendid building, vacant only since 1969, may yet escape demolition by being transported to another site.

The George Sealy carriage house and stable in Galveston, designed about 1890 by McKim, Mead and White; now owned by the University of Texas Medical School.

The elegant bank-barn carriage house at Terrace Hill, Des Moines, Iowa. Designed in 1869 by William W. Boyington, this Second Empire hilltop edifice, now the state executive mansion, was the residence of the fabulously wealthy P. M. Hubbell, who established a trust fund for its maintenance.

While Belcourt was under construction, Hunt was also occupied with his most ambitious Newport scheme, the Breakers. Modeled after a Genoese palazzo, the Breakers is the largest, grandest, and most successful of Hunt's Newport cottages, built for Cornelius Vanderbilt, Sr., grandson of the Commodore. The stables of the Breakers are not attached to the house. In fact, they are not on the grounds at all, but it would be a mistake to think that the Vanderbilts had less interest in their horses than did Oliver Belmont, even if Cornelius did set his stables a little farther from the hearth. From the Commodore's time on, the Vanderbilt family has been associated with horses in one way or another. In 1865 the Commodore, driving himself in a high-wheeled racing cart, won a match race against the aging but still formidable Flora Temple with his big chestnut, Dan Mace. The occasion was a gala afternoon of trotting races in honor of Ulysses S. Grant. The Commodore's son, the first William H. Vanderbilt, had a passion for trotters and carriage horses, which were housed in a palatial stable, now vanished, at Madison Avenue and Fifty-second Street. For six years he owned the record-breaking trotting mare Maud S., using her as a road horse and winning many an informal "brush" with his friends. Maud S.'s record for the mile stood for many years; it was broken in 1891 by Sunol, a daughter of Leland Stanford's Electioneer. Most of Vanderbilt's extensive collection of vehicles was eventually transferred to his son's coach room in Newport.

As it stands today, the Breakers stable is a massive one-story brick block with twin carriage doors, one of which opens into the large garage. The frame

98

second story, which contained stable manager's and chauffeur's apartments as well as the loft, was destroyed by arson in recent years. It was built around a big central skylight over the carriage room. The first floor, now receiled and roofed in, came through the fire relatively unscathed, but many of the irreplaceable vehicles on display were badly damaged and traces of charring may still be seen in the fabric of the building.

In spite of its scale, this is an extremely efficient layout. The stall range is like that at Belcourt, and runs across the rear of the building beneath arched windows. The flooring in the stable area is small raised brick, laid over a sophisticated system of aisle and stall drains. As at Belcourt, the aisle floor was

The Breakers stable, Newport, before the fire.

sanded, and the stalls "set fair" with straw each day. Each straight stall has two brass tie rings, so that a bucket may be suspended there; there is no watering trough. The aisle is served by two wash closets with water laid on, two hay drops with ladders, three sealed feed chutes, and a manure trap at either end. (The greenhouses are not far below the stable.) The corridor from stable to carriage entrance separates the carriage room and the harness room behind it from tack room, staff locker room, dispensary, and office. Sliding glazed doors cut off the carriage room from dust or stable odors, and the stall range may also be closed off in cold weather. Stall drains, latches, fold-away robe rails, turned walnut whip reels, and the like came from the Mott Iron Works in Brooklyn, source of some of the loveliest stable fittings of the period.

Like Belcourt, the Breakers is open to the public during the summer months, and before the fire the stable and its museum collection of equipages were also on view. Happily, among the vehicles to escape the flames was Alfred G. Vanderbilt's road coach, *Venture*. Born in 1877 and perishing in the torpedoed *Lusitania* in 1915, Mr. Vanderbilt was an internationally respected

Interior of the Breakers stable, with the stalls "set fair."

The stable of noted whip H. H. Rogers, Fairhaven, Massachusetts, designed by Charles Brigham. The length and quality of the straw with which the stalls are set fair indicates the age of the photograph.

whip. At one time he drove *Venture* on a regular schedule between New York and the old Ardsley Club (demolished about the time of the Second World War) up the Hudson, and later revived a London to Brighton run. His matched grays were famous; his son, Governor Vanderbilt, kept up the tradition. When the American Carriage Association lately met at Newport and horses again filled all twenty-eight stalls, a contemporary exponent of the art of four-in-hand driving tooled *Venture* behind his own gray team through the streets of Newport. As if time had rolled back, an elderly man at the curb broke off his conversation to call out, without visible surprise, "'Morning, Governor!"

It is to be hoped that this building can be restored and reopened. Apart from the still notable assemblage of nineteenth-century vehicles within, it is a model of intelligent design, well-suited to its original use, and, despite its size, not without lessons for today.

One of the first William H. Vanderbilt's daughters married Elliott F. Shepherd, and although in New York they shared her father's Fifth Avenue mansion (razed in the 1940s), they also enjoyed a country estate in Scarborough-on-Hud-

101

Henri de Lattre. The Stuyvesant Family Stables. *Private Collection.*

son. Woodlea, now the Sleepy Hollow Country Club, was built in 1893 by Stanford White. There are a number of houses and outbuildings on the place, including an inviting red-brick courtyard stable in the Norman style, converted to other uses years ago; but the twin gatehouses, the mansion, and the main stable are all of buff brick trimmed with sandstone, drawn up in White's inimitable blend of Italian Renaissance and neoclassical Georgian. The main house is very sizable indeed; it contains sixty-five rooms, exclusive of domestic offices, on its three floors, and none of them is small. The stable is on an appropriate scale. It is even larger than it appears, as advantage has been taken of the slope of the ground to put in additional stalls below. For many years it has been operated by the Sleepy Hollow Country Club as a semipublic riding establishment, and there are no longer any straight stalls on either level, but originally there were sixteen box stalls and sixteen straight stalls on the upper level and twenty-eight straight stalls downstairs. Even today, with every horse in a box, there is

102

accommodation for fifty: an extraordinary capacity for a private stable, but an excellent arrangement for a semicommercial one. The upper ranges are lit by skylights, and there is a spacious mow. An indoor arena, open to the roof, separates carriage house and stable block. The huge tack room, which once held long cleaning racks and endless rows of harness and tack supports, has been streamlined. The brick of the upper floor has been covered with asphalt, and an automatic loader has been installed in the loft. None of the changes has destroyed the impact of the building, and behind its somewhat pretentious facade, it works very well.

Large, creative, and luxurious carriage house–stables were being built elsewhere in the United States in those days, if generally in less overwhelming modes. The architect of the stable building supporting Belle Meade, in Nashville,

The Sleepy Hollow Country Club stables, Scarborough-on-Hudson, New York. The central block is a covered manège, open to the roof.

103

Tennessee, is not known (color plate 13). It was built circa 1890 by General William Hicks Jackson. The Belle Meade Stud, established by the Harding family and brought into international prominence by their distinguished descendant General Jackson, compiled an enviable record. Among the animals owned or bred at Belle Meade were Enquirer, great-grandsire of Man o' War; Luke Blackburn (by Bonnie Scotland out of a Lexington mare — Sir Archie's line running true), who won twenty-two out of twenty-four starts as a three-year-old; the money-maker Bramble; and Iroquois, the first American horse to win the Epsom Derby, which he did under Pierre Lorillard's colors in 1881. Iroquois went on to win the St. Leger the same year, a feat unmatched by an American entry until Never Say Die's victory in 1954. The old stallion ended his days at Belle Meade, and is buried not far from the appealing antebellum house. Perhaps, as a pensioner, he occupied one of the stalls in the coach barn.

This frame building is T-shaped, with the stable in the crossbar. Stable and

104

The vanished Flagler house, Park Avenue, New York City, and (right) its carriage house around the corner at 149 East Thirty-eighth Street. The Dutch Revival mansion has been demolished, but the stable building survives.

carriage house share one roof and a continuous upper story, but an open breezeway separates them at ground level. The stable itself was floored with peat moss and clay; a large central space is surrounded by huge box stalls, interrupted by two pairs of straight stalls and additional doorways. The loose-boxes need their extra size; perhaps used for harness horses with set tails, their interiors are curiously wainscoted on three sides with sloped and curved buffers of narrow barrel staving set at an angle of about sixty degrees, so that they rather resemble old-fashioned chocolate cups. These buffers may have been an elaborate substitute for tailboards. In any case, they are certainly out of the common way, and if their value is moot, still the man who put them in certainly knew his way around a horse. There are fourteen stalls in all, each with a barred window. Over the stable is a mow with centered hay drop, and there are seven grooms' chambers (or bachelor quarters) over the carriage house, with additional loft space above. The louvered eyebrows set into the roof, the airy cupola, and the stick-style trim on the dormers combine to lighten the generally austere lines of the building. A shed with hitching bar runs along the near side of the carriage house, offering temporary shelter for visitors' horses; perhaps their vehicles were drawn inside. The breeding and training barns at Belle Meade have disappeared, for the plantation was divided and sold off upon Jackson's death in 1903. The house and its outbuildings are now the property of the Association for the Preservation of Tennessee Antiquities, and are open to the public.

There is a highly unusual group of six nineteenth-century stable buildings near Gallatin, Tennessee. Each stall, at one time surrounded by its individual paddock, is a freestanding brick block, complete with Dutch door, small access door, and loft. They were built to house stallions; set along the entrance drive of the plantation, they allowed prospective clients to observe the available sires for themselves, before meeting the owner or manager or examining the young stock. These were Thoroughbreds, and beyond the lovely old house stands a vintage training barn of a pattern familiar in the South from the Mississippi to the eastern seaboard. The double row of stalls, broken by a cross-aisle with adjacent feed, tack, and equipment storage, is covered by a low-pitched roof extending over a fenced aisle on the perimeter of the block. Here green and skittish babies may have their early lessons or warm up for more advanced exercise in a safe and familiar confined area.

Breeding and training farms all over the country were building enormous and inventive facilities in various styles of architecture toward the end of the nineteenth century; unfortunately, most of them survive only in pictures or in

Individual brick "stables" near Gallatin, Tennessee.

106

YOU ARE SPECIALLY INVITED TO VISIT THE MOST FAMOUS HARNESS HORSE BREEDING FARM IN THE WORLD, THE HOME OF THE WORLD CHAMPIONS, DAN PATCH 1:55   CRESCEUS 2:02¼, DIRECTUM 2:05¼, ARION 2:07¾.

The main stable was designed by M. W. Savage with a view of obtaining good ventilation, sunshine and fresh air in every box stall. We believe this to be the only stable of this kind ever built. The octagon center is 90 feet in diameter and each of the five wings is 157 feet long and they contain 130 box stalls. The center is over 100 feet high and contains a large water tank in the top of the dome which gives a water supply all over the stable. This tank is filled from a large spring near the stable which has a flow of 5,000 barrels per day. The entire stable is heated with steam and hot water. We use only the hot water system for the horse stalls, as this gives an even temperature. The octagon center is floored with cinders and in this we "bit" our weanlings. The stable also contains sleeping, reading and bath rooms for the men and was erected at a cost of over $50,000. Beyond the stable can be seen the mile track built by the famous track builder, Mr. Seth Griffin, at a cost of $18,000. This is one of the best mile tracks ever built and here is where we train the colts sired by our champion stallions, Dan Patch 1:55   Cresceus 2:02¼, Directum 2:05¼ and Arion 2:07¾. To the left of the stable is located our half mile track which is for use when the mile track is too heavy from rain. In this way we can train our colts every day regardless of weather. In the distance can be seen the high bluff upon which is located the summer home of Mr. Savage. At the foot of the bluff flows the beautiful Minnesota River on its way to join the Mississippi at old historic Fort Snelling, twenty miles to the east.

One of Mr. Savage's advertisements.

memory. Marion Willis Savage's International Stock Food Farm was certainly one of the most interesting and even flamboyant of these mammoth creations. Embellished with Roman arches, ocular windows, cupolas in the shape of minarets, and focused on an octagonal hall (almost as large in itself as the Mount Weske stable) crowned by an imposing dome that doubled as a water tower, the composition reduces its extensive and workable stable areas to the status of mere connectors. The plan has real advantages beyond those set forth in the advertisement shown here; indeed, perhaps the designer would not have had to boast of his heated stalls, at best a questionable amenity, had not his barns been as exposed to Arctic winter blasts as to mild "sunshine and fresh air." But Savage could well afford to coddle his charges. His International Stock Food Company was doing very well on its own merits in 1902 — well enough to allow "The Parson," as he was called, to pay sixty thousand dollars for Dan Patch after the legendary pacer broke the two-minute mark for the mile. The rest was show business — and history.

In his private railway car, Dan Patch traveled all over the country, pacing against the clock. He earned substantial sums for each appearance, brought in more than fifteen thousand dollars a year in stud fees, and incalculable values in advertising and public relations. His name was attached to hundreds of products, from toys to washing machines; it became literally a household word. The bay wonder went from strength to strength on the track, although he competed only against time. In 1906 he achieved his unofficial record of 1:55 at Hamline, Minnesota. In 1909, the holder of nine world records, he retired, thereafter only making occasional exhibition trips with some of his famous stablemates. He died on the eleventh of July, 1916. M. W. Savage died on the twelfth.

The old training barns for Standardbreds are of a special pattern, one that has been adopted and adapted by American Saddlebred and Tennessee Walker farms. Like the Leland Stanford barn, they were long and contained an exercise area between the stall ranges: an aisle wide enough to allow the horse to turn with a following vehicle. Certainly the largest of these to survive, and in its own severe way perhaps the handsomest, is the Long Barn of Walnut Hall Stud, now the Kentucky Horse Park, in the Bluegrass region of Kentucky; it is four hundred and sixty feet long and sixty feet in width.

Walnut Hall has an interesting history. Christened in 1816, by 1830 it was a thriving Thoroughbred farm and continued as such under various colorful owners until 1892, when it was bought by Lamon V. Harkness of the Standard Oil Company. It is said that Mr. Harkness first came to Lexington to look at a team of French Coach Horses, and like many another before and since, fell in love with the Bluegrass country. Be that as it may, Mr. Harkness was a trotting-horse man, and Walnut Hall became the home of distinguished Standardbreds. In ten years he expanded his acreage from four hundred and fifty to two thousand acres, and his broodmare band from twelve to two hundred, standing such outstanding sires as Allie Wilkes and The Harvester. In Mr. Harkness's day, the entire yearling crop of the farm was sold every year at auction, a great boon to the industry. He also paid great attention to such practical matters as water supply to his pastures, suitable feeds, and appropriate stabling. After his death in 1915, he was succeeded by his daughter and son-in-law, Dr. and Mrs. Ogden M. Edwards, Jr., who brought in or bred at Walnut Hall a long list of great sires, including Guy Axworthy, Peter Volo, and Volomite. This high standard is maintained by Mr. and Mrs. H. Willis Nichols, Jr., at Walnut Hall Farm.

The Long Barn was built in 1897. Thirty-eight double box stalls, interrupted by a raised observation stand, line the wide dirt "aisle." Twin open lofts ceil the range, their wainscot balustrades punctuated by handsomely trimmed door-gates for dropping hay and bedding; the center is open to the roof. Simple, solid, and spacious, its noble proportions reconcile the eye to its size. It was built to be used as a stallion barn in the breeding season, and as a bitting and sales barn for yearlings.

Walnut Hall Stud and Walnut Hall Farm were divided after World War II, and in 1972 the Stud was purchased by the Commonwealth as the site for the Kentucky Horse Park, which formally opened in 1978. The last of the harness racing "Immortals" to stand at Walnut Hall Stud was Rodney, who is buried at the end of the Long Barn, now restored as a part of the Horse Park. Rehabilitation has been confined to preserving the fabric and introducing such conveniences as electricity, running water, grooms' facilities, and the like. Sixty temporary stalls in a double row may be set up down the center, leaving ample aisle space to either side; more than ninety horses may be comfortably housed at one time. Thanks to the 1978 World Championship Three-Day Horse Trials, the Long Barn now shares with Badminton an illustrious roster of Event horses to which it has given shelter.

The Long Barn during the 1978 World Championship Trials. See also color plate 17.

Octagonal stallion barn, now a machine shed for the Harmony Landing Country Club, Louisville, Kentucky. It appears to have been modeled after James Ben Ali Haggin's stallion barn, but on a larger scale.

The Kentucky Horse Park is a farsighted venture. It supplies a capsule education in the history of the horse, particularly the horse in Kentucky; it provides, through the dedication of the Park staff and the Director and personnel of the Equestrian Events Committee, a superb facility for horse trials and other equine-related activities; and it conducts a practical training program for those seeking employment in the horse industry. The original suggestion for the Park came from Lexington horseman John R. Gaines, in the cause of those who come from a distance to see the Bluegrass region for themselves, only to find the great breeding farms closed to them. The ever-increasing number of visitors, the chronic shortage of trained staff, and the rising incidence of carelessness and vandalism have caused owners and managers to rethink their old hospitable ways, but the Park is meant for visitors and welcomes all who come.

While continuing a family interest in Standardbreds that goes back to the 1890s, John Gaines moved into the Thoroughbred business in 1962. In 1966 he purchased the present Gainesway Farm from C. V. Whitney to accommodate his expanding and already prestigious nursery; in 1975 the farm was described by a London publication as "one of the most successful breeding operations in America." Money, inspiration, and old-fashioned horse sense have gone into its development.

The year 1966 was an auspicious one for Mr. Gaines. He bought his new farm, leased and syndicated the fabulous Bold Bidder (which won three of the richest races in the country for him that season before retiring to stud, breaking one long-established record and straining another in the process), and he saw his homebred Kerry Way win the historic Hambletonian in record-breaking straight heats. The site of the new Gainesway was itself auspicious, for fine horses have flourished on that land since before the turn of the century, when it was still part of James Ben Ali Haggin's vast Elmendorf Stud.

110

Born in Kentucky, Haggin (1822–1914) made millions out of the mines of the Far West, and established in California the largest Thoroughbred breeding operation ever recorded, for he was a man who "thought big" in all his dealings. But the Rancho del Paso was a long way from the sales barns of the East, and he eventually transferred his horses to his native state. Elmendorf in its prime measured eight thousand seven hundred acres of the best of the Bluegrass. One of the properties included in the spread had belonged to the man who sold Haggin his most successful racehorse, Salvator, which, with the great black Kentucky jockey Ike Murphy in the irons, achieved many victories and much popular acclaim. Isaac Murphy is now buried near Man o' War at the Kentucky Horse Park.

The greatest runner Haggin bred himself was Firenze, that early triumphant exponent of female equality, who raced against the "stronger sex" for four years and was unplaced in only five of eighty-two starts. Haggin is chiefly remembered, however, for the enormous numbers of top-quality yearlings he bred and sold from the endless pastures of Rancho del Paso and Elmendorf. The modern Elmendorf Farm, Greentree Stud, Normandy Farm, part of Spendthrift, the C. V. Whitney Farm, and the adjoining Gainesway were all once part of the original Elmendorf Stud.

Haggin liked polygonal buildings. He built an octagonal stallion barn, now used for storage, and twin twelve-sided stables, one of which serves as a yearling barn at Gainesway. The plan of this building is extremely simple; a single

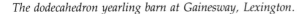

The dodecahedron yearling barn at Gainesway, Lexington.

faceted ring of generous box stalls, three to a side, encloses an open grass plot. Every stall has a door and window to the outside, and a second set of openings under the inner shed roof. Animals may be moved from stall to pasture or into the enclosed central area, and the vehicle access may be closed off at will. Stall doors are solid board and batten, hinged against the wall, with a mesh stall-guard or drop bar to restrict the tenant when they are open. There are slots at the base of the doorframe to hold broad plank draft stoppers; every aspect of ventilation is controlled. The building is frame, of vertical boarding roofed with tin, and the stalls are clay-floored. There is a firm track under the shedrow where feed and manure carts may pass. Except in the worst of weather, when in any case yearlings are unlikely to be in residence, this must be an outstandingly convenient barn to tend, as well as offering creature comforts and the companionship of neighbors to the eager adolescents stabled there.

There is a formal stone and brick domestic stable in Maryland, almost as far removed in design and finish from the building discussed above as may be imagined; but today, revived as a museum, it commemorates a tradition extending over more than two hundred years of American Thoroughbred history. It is the carriage house and stable built by James Woodward at Belair, Prince Georges County, at the end of the nineteenth century, the last successor to a series of substantial stable buildings that may be traced back to 1769. The men associated with the early days of the plantation — Samuel Ogle, his son Benjamin, and the Benjamin Taskers, father and son — all played a part in establishing the foundation Thoroughbred bloodlines in this country and in fostering the sport of racing in the New World.

In the first half of the twentieth century, William Woodward (1876–1953) re-established Belair Stud as a force to be reckoned with on the turf and in the sales ring. When Mr. Woodward inherited the estate from his uncle in 1910, he rebuilt the frame stable wings in brick. They housed his carriage horses, hacks, and hunters, and later sheltered yearlings, in from Claiborne in Kentucky (for Woodward worked closely with A. B. Hancock of Ellerslie Stud and Claiborne Farm) to be chosen for sale or training. Like John Gaines a generation later, he held that nothing is more important to a breeding farm than the "quality of its matrons" and early established an impressive group of mares, many imported from war-torn France. Nevertheless, he stated in after years that it had taken him twenty-five years to achieve the broodmare band of his dreams.

Mr. Woodward joined with A. B. Hancock and others in syndicating the French stallion *Sir Gallahad III, who sired Gallant Fox on a Belair "matron" in his first American crop of foals. Gallant Fox, only the second horse to sweep the Kentucky Derby, the Preakness, and the Belmont Stakes, brought home the Triple Crown in 1930. At stud, he sired Omaha, who repeated his father's triumph in 1935. The list of running greats, born at Claiborne and brought to Belair to be readied for sale or training, is a long and honorable one. The last of these was Nashua, who finished his racing career in the colors of Leslie Combs

The Belair stable, Prince Georges County, Maryland. In this view of the entrance front, one of the stall wings may be seen beyond the arch.

II of Spendthrift Farm, head of the syndicate that purchased him following the untimely death of William Woodward, Jr., in 1955. *Nasrullah, Nashua's great sire, fathered a line of champions.

An interesting footnote to Belair's Thoroughbred history involves the road coach *Pioneer*. It briefly occupied the Woodward carriage house after a 1916 Coaching Club expedition from New York, taken by stages with a change of horses at every halt. A clean-bred team supplied by William Woodward drew the vehicle — weighing five thousand, three hundred pounds under load — on the last stage from Baltimore to Belair; an emphatic demonstration of the strength and versatility of the Thoroughbred. It must also be mentioned that Woodward, who found profit as well as pleasure in his horses, paid tribute to the great race mare Selima (who lived at Belair in Samuel Ogle's time and who made a considerable mark on American foundation stock) in the form of various perpetual trophies established in her honor, and a plaque which is still affixed to the Belair stable wall. The Belair Stable Museum is operated by the Bowie Heritage Committee and is open to the public.

The Belair stable survives only as a symbol, victim of personal tragedy and a changing society. Ben Ali Haggin's dodecahedrons are still in use because of their functional construction and performance; their aesthetic success has always been incidental. The studied elegance of the contemporaneous brick stableyard at Nydrie, in Albemarle County, Virginia, on the other hand, was of the first importance to its contriver and still makes an important contribution to that flourishing Thoroughbred nursery. Built as a carriage house and stable to complement Harry Douglas Forsyth's Scottish-baronial seat, it now provides Nydrie Farm and Stud, a progressive and streamlined operation, with central offices, feed and equipment storage, an eminently satisfactory yearling barn, and an unforgettable image.

113

The courtyard at Nydrie, Albemarle County, Virginia, which achieved its final exterior form in 1933.

The courtyard view of the carriage block, Nydrie.

114

The connected buildings are laid out in a closed quadrangle, pierced by arched entries with massive double door-gates that pass through gabled lodges. The warm rose brick of the walls was manufactured on the place, and the mill-work, including the stall interiors, which have heavy wainscots with bull-nose returns, is oak. The lofty carriage house, capped with a stepped cupola and weather vane, projects into the square; the coach access and hay door on the courtyard side are repeated on the outer wall. Farm offices now occupy the old coachman's or stud groom's quarters to the left of the carriage house. The Stud offices are in the big stalls that butted them; the oak liners have been retained and brought to a glowing finish. Opposite the carriage house are twenty-one shedrow loose-boxes, and there are two foaling stalls in the link which corresponds to the present Stud offices.

The architectural style of the scheme is a trifle elusive — a touch of the Cotswolds, a suggestion of the Tudor — but achieves a charm and integrity wholly satisfying to the eye. The effect is heightened by immaculately raked soft gravel, the green precision of box hedges, and splashes of color in flower boxes and jardinières.

Since 1927 Nydrie has been owned by the Van Clief family. Mr. and Mrs. Ray Van Clief restored the great main house and moved to Virginia, looking forward to a healthy and secluded country life. In 1932, Mrs. Van Clief bought six broodmares with foals at foot from A. B. Hancock, Sr. The well-bred foals were sold at the 1933 Saratoga sales, and Nydrie Stud became a going concern. Daniel Van Clief took over its management in 1952. The strength of the farm has always been in its splendid broodmares, bred to carefully selected sires; Nydrie does not stand a stallion. Between ten and twenty yearlings travel to Saratoga from Nydrie each year, and as the Van Cliefs have four grown sons, more than one of whom plays a role in Farm and Stud affairs, the future of Nydrie seems secure.

Mr. Forsyth's castellated mansion, a painstaking reproduction of his ancestral home in far-off Sterling, molders silently among its abandoned plantings. Too difficult to maintain and too monumental for demolition, it was, in the end, regretfully abandoned by the Van Clief family and is gradually succumbing to weather and the probing fingers of roots and vines. The lower windows and the terrace door are gagged and blindfolded with ivy, and a nut tree has rooted in the cellar and thrust its way through floors and ceilings to burst triumphantly into leaf at the eaves. Built out of its time, the somber pile has already acquired, like the ruins of some prehistoric fortress, an air of haunted unreality. But at the stableyard, the nerve center of today's Nydrie, the buildings are perfectly at home in the present. Here the ivy is an ornament, not a threat, and in June and July the curious yearlings peer out over their stall-guards at the purposeful activity in the yard. While the gloss that they will carry to Saratoga in August is preserved and enhanced with meticulous care, they spend their days developing a comforting familiarity with the busy outside world.

The crumbling turrets of the mansion at Nydrie are emblematic of a curious and not uncommon atavistic impulse. The great eighteenth-century plantations of the Southeast were logical reflections of the European architectural estates on which they were modeled; their self-sufficiency was a necessity in what was still, in spite of graceful flourishes, a pioneer society. The gargantuan nineteenth-century domains of Stanford, Baldwin, and Haggin, on the other hand, were essentially part of a romantic revival; created at a stroke out of wealth flowing from other sources in order to satisfy a yearning to be, in the oldest and most complete sense, landed. This impulse partly underlies the homesteading movements of today, as well as modern efforts to re-establish the viable family or communal farm, but perhaps the most outrageously ambitious versions of these private empires were founded in the early 1900s by two Ohio industrialists.

In 1903, Dr. Samuel Hartman began to pour the vast profits derived from his patented "elixir," Peruna, into a stock farm in Franklin County. He constructed a private electric railway to carry passengers and freight between Columbus and the farm, put in his own power plant and water supply, erected employee housing, a little red (brick) schoolhouse, and a hotel, and raised a series of huge frame barns to house, along with other prized animals, the world's largest herd of registered Jerseys, which at one time supplied milk to the whole city of Columbus. By 1906 the farm was the largest of its kind in the world. Hartman bred three strains of horses, founded on imported stock, which were all of championship caliber; his intriguing choices were Arabians (of Polish blood), Percherons, and German Coach Horses.

The two great frame horse barns that are still standing were built in a style familiar throughout central Ohio, but are distinguished by their elegant Gothic louvers and their stupendous size (color plate 16). They measure one hundred and forty by sixty feet and contain twenty loose-boxes and forty tie stalls; all their fittings were supplied by the Mott Iron Works. The 1905 farm catalogue states that the "Hay Mow Capacity is 800 Tons." They are superb buildings, both handsome and utilitarian, and splendid examples of load-bearing frame construction; but at this writing it seems sadly unlikely that they will be preserved or restored, for the property is intended for development.

A hundred miles to the north, just outside Akron, five remarkable buildings, erected between 1909 and 1915, still bear mute witness to the most visionary of these little kingdoms. Ohio C. Barber, the Ohio Match tycoon, having been to see the Hartman stock farm, was inspired to emulate and surpass it. Anna Dean Farm, named for his adored wife, once covered thirty-five hundred acres. The farm was entirely surrounded by post-and-rail fencing of reinforced cast concrete, for Mr. Barber was convinced of the superiority of masonry construction for every purpose, citing its impressive appearance, durability, and immunity to fire. Men, beasts, and machinery were all housed in steel-framed brick and concrete-block buildings, roofed in red tile and executed in a ponder-

116

One of the remaining horse barns at Hartman Farm, Franklin County, Ohio, in an early photograph. See also color plate 16.

ous Second Renaissance Revival idiom; as at Santa Anita, everything required for human and animal sustenance was raised on the place.

Mr. Barber had a touching faith in the rightness of all his works. He wrote a monograph on the glories of Anna Dean Farm, extolling (in the third person) the genius of its conception, the beauty and efficiency of its operation, and the unparalleled quality of its livestock. He experimented extensively with gasoline and steam-driven equipment, but still felt that on a farm "there remains a demand for reliable, light, easily handled power that only the horse can supply." He rode his acres on fine Saddlebreds, had upward of one hundred and fifty draft horses in work, and maintained a stud of registered Belgians with a view to upgrading and eventually replacing his more plebeian teams. The workhorses were hitched in threes, "eliminating the need of two drivers for every twelve horses," and thus theoretically achieving better and faster results.

The big horse barn, where the straight stalls of the working teams and the loose-boxes for the Belgian broodmares were, has been demolished, as has the stallion barn across the way, where every stall opened into its own massively concrete-fenced paddock. Of the original stables, only the comparatively modest "Colt Barn" still stands. The stalls are of parged cement block, with concrete

117

The vast, vanished work-horse barn at Anna Dean Farm, near Akron, Ohio, 1910.

floors and metal doors barred like stanchion gates; even the mangers are cast concrete. Hay and bedding were stored in the fireproof loft. The building was lit by electricity and supplied with metal litter carriers and feed wagons. It had its own heating system, which must have been needed not only in winter but in periods of prolonged humidity. The stable was the home of the pampered get of Jupiter Chief, mighty herd sire, who brought home state fair championships with monotonous regularity; his name may be found in many a modern Belgian's pedigree.

So successful did this horse-breeding project become that the grandiose 1915 Piggery, known as the Pig Palace, was eventually converted for stabling. It has since been briefly used as a dairy and a residence, for when Mr. Barber died, the farm fell into disuse. Of the hilltop house from which he viewed his lands nothing remains; even the cellar hole has been filled in. The Piggery and Colt Barn are the last architectural vestiges of what truly was, in its brief hour of grandeur, "the world's greatest farm."

The farm village at Vizcaya, James Deering's winter home in Miami, was, comparatively speaking, a modest effort, a charming but practical amenity which supplied milk, eggs, poultry, and vegetables to the house and included a shrub and flower nursery for the ten acres of formal Italian Renaissance gardens. The farm is focused on a baroque North Italian-style hamlet, an extraordinarily successful exercise in the sophisticated use of a vernacular style, and the

118

The Colt Barn at Anna Dean Farm, 1911.

logistical heart of the estate. Concealed behind its romantic facades were paint and carpentry shops, a forge, the garage and machine shop, a laundry, extensive staff quarters, a carriage house, poultry coop, dairy, and stable. For many years the village has been the headquarters of the Metropolitan Dade County Parks and Recreation Department, and the interiors have consequently been much modified; but the two-story stable, its cupola ventilator disguised as the cap of a wide central chimney, and its utilitarian hay and stable doors recessed behind a pierced brick wall, retains an air of almost prim elegance. In Mr. Deering's time this little cameo of a building housed the mules that tilled the fields and gardens of the farm; their plows, rakes, and wagons were kept in the adjoining carriage house, the steep gable of which complements the flat stable roof. Mules can rarely have enjoyed so notable a dwelling.

This fantasy farm enclave was designed, like the palazzo it supported, by F. Burrall Hoffman, Jr., in collaboration with Paul Chalfin. Vizcaya itself was loosely modeled on the Villa Rezzonico (ca. 1670), which stands in the Brenta River country north of Venice, and it successfully incorporates the extensive collection of beautiful but unrelated architectural fragments acquired by Mr. Deering after his retirement from the International Harvester Company. The

119

The bijou 1916 mule barn at Vizcaya, Dade County, Florida. The pitched-roof building is the carriage house.

formal gardens, the inspired achievement of Diego Suarez, are a superb setting for the fountains, balustrades, pillars, and statuary brought from Italy.

Vizcaya with its pleasure grounds, now the Dade County Art Museum, is, except for Biltmore, perhaps the best-known and most successful of American palaces. It served its purpose for the builder, who took most of his joy in it from the ten years of its creation, and today it gives pleasure to thousands of visitors who annually benefit from the generosity of his heirs. It now seems probable that the village across the way will be acquired by the Museum, so that visitors may once more see the estate as the unified whole it was designed to be.

120

VII *A Way of Life*

The pick of the basket, the best in the shop,
Is the clipper that stands in the stall at the top.
—JOHN WHYTE-MELVILLE

Major Whyte-Melville, who wrote so stirringly of horse and hound, died, as he would no doubt have chosen to do, of a fall in the hunting field in 1878. So many of the stables discussed in this chapter (built, broadly speaking, between the First and Second World Wars), are dedicated to the horse as a sporting companion that it seems fitting to quote that gallant Georgian horseman at the head of it.

Much has been said and written about the American Establishment, most of it after the fact. In the early decades of the twentieth century, however, there really was a comfortable loose-knit order of families of adequate means who subscribed to the same conventions, often attended the same schools, and generally shared the same, largely sporting, interests. Then, as now, golf, tennis, and other racquet sports had their adherents, and, of course, there were those never happier than when "messing about in boats," but they were also, in fact, a truly "horsey" set. They founded or expanded stud farms and racing stables, played polo with verve and skill, continued to drive their smart turnouts in the show ring when motor traffic forced them from the roads, and rode, often on homebreds, to hounds and in hunt races. Although sleek and powerful automobiles brought them up — or down — from "town," stables were as vital as garages to their country homes. "Their lines had fallen in pleasant places," and there is an air of relaxed self-assurance about their residential buildings that reflects the confidence of owners and architects in the permanence of that vanished world.

There are certain areas in the United States where the horse population has been from time to time so concentrated and cultivated that a book could, and probably should, be written on the stable buildings of each of them; their heyday was between the wars. Among these are Long Island, the North Shore above Boston, the Bluegrass of Kentucky, the piedmont of Virginia, the Carolina midlands, the Santa Ynez Valley in California, and the Florida highlands around Ocala. All these areas have much to show the stable buff; of them all, Long Island has the longest history as a center for equine activity.

The first American horserace on a planned and measured course was run on the Island in 1665. The Meadowbrook Hunt, founded in 1877 and claiming descent from the Brooklyn Hunt, established in 1744, was attracting fields of two hundred or more in the 1890s; by 1910 the Smithtown and Suffolk hunts had been added to the roster. The Suffolk was relatively short-lived, and in the 1970s the Meadowbrook country dwindled to such an extent that it was forced to suspend operation, after showing sport for a century. One by one the farms and estates have been swallowed up by urban and suburban development, yet houses and outbuildings of a more spacious day are still to be seen.

Thomas Hastings (1860–1929) was a founding partner in the prolific and influential firm of Carrère and Hastings, creator, among other works, of the New York Public Library and the House and Senate Office Buildings in Washington. Hastings himself is also identified with a number of noted private homes, including the Flagler house in Palm Beach and the Henry C. Frick house in New York, both now museums. Mr. Hastings built a family residence for himself in Old Westbury, in the Meadowbrook country, called La Bagatelle, to serve as summer home and hunting box (color plate 19). Completed in 1912, and partially rebuilt in 1916 after a destructive fire, it follows, in plan, the architect's metaphorical dictum "one half for the pudding, the other for the sauce." The house is approached through a courtyard. Directly opposite the gate is the carriage house; the six Dutch stall doors of the stable face the north or entrance front of the modest mansion. Save for the snowy marble of the entrance loggia and the fountain, the exteriors are of dark, hard-burned brick. Behind the carriage house are the tack and feed rooms; and the well-lit aisle behind the stalls gives access to quarters for groom and chauffeur and to a rear garage yard with washrack and watering trough. It is a delightful and ingenious scheme. The terrace and informal wooded garden are on the south and west sides of the house and court, walled off from any sight of the service areas, yet as the owners passed through the entrance hall they could glance out at his hunters or her driving horses (for Mrs. Hastings was a founding member of the Dorcas Coaching Club, or Ladies Four-in-Hand Driving Club, as it was officially described). The family wintered in South Carolina in another owner-designed house (page 206) which neatly incorporated the stable into its composition.

Mr. Pierre Stevens, the present owner of the Old Westbury property, is in the process of restoring house and surrounding acreage to their former state. The stable wants very little to bring it into first-class condition; structure and stall fittings are sound. Mounted on the rear walls of the generous loose-boxes are stall plates commemorating favorite horses. Holdfast, by Alloway out of a mare called Squeeze, was one, and Cherokee, "bred at Biltmore — bought at

Aiken," another. Cherokee died in 1928. A year later, his owner succumbed to appendicitis, but the memorials to his confidential hunters remain.

The Lathrop Brown estate at St. James, now the Knox School, is the site of a particularly interesting stable, the only one of its kind known to have been built on Long Island. It takes the form of a horseshoe; a bar shoe, in this case, as the heel flanges — one a stud groom's cottage and the other holding harness, carriage, tack and trophy rooms — are connected by a hyphen gateway pierced by a pedimented arch. In addition to its symbolic appeal, this virtually circular configuration is highly advantageous, combining the easy observation and maintenance of the Gainesway dodecahedron with immediate proximity to essential support systems. This may have been the stable to which James C. Mackenzie refers in *Sporting Stables and Kennels*, which he published with Richard V. N. Gambrill in 1935, although his illustration "in the French manner" differs stylistically from the shingled Queen Anne idiom of Peabody, Wilson and Brown, who were responsible for the Knox School building, and is a trifle smaller.

The central door-gate is the only major access to the grassed interior; the dramatic shingled water tower rises beyond the stable on axis with the arch, but the roof over the stalls is uninterrupted. The building has been well maintained, and still has horses in it, as well as providing additional classroom space where a feed room and two stalls adjoining the cottage have been thrown together and fully enclosed. The curve of the walls is noticeable in this extended space, but not in the stalls themselves.

Side view of the Lathrop Brown stable, now a part of the Knox School, St. James, New York; before restoration.

Overleaf: The Sundridge yearling barn, Upperville, Virginia. See also color plate 22.

123

The circular form has always attracted designers and builders, and has been variously justified and employed. Jeremy Bentham's 1781 "Panopticon" plan for a prison to be built on "enlightened principles," shows what is in effect an immense, open, multistoried polygonal or circular barn with cells replacing stalls on the perimeter.

The semicircle has also had its design advocates, particularly among adherents of the Palladian school. The yearling barn at Sundridge, part of Paul Mellon's Rokeby estate in northern Virginia, is a particularly handsome example. The plan echoes Tryon's Palace or Mount Airy (pages 40, 36); the tall central block with its triple-bay projecting loggia is connected by curved nine-stall ranges to matching two-story flankers, forming a demilune. We have been unable to discover the designer of this novel and charming composition, but on-site

The curved stall ranges at the Sundridge barn.

The stable of William J. Creighton's design at Llangollen Farm, Upperville. Opening to the rose garden and the house beyond, the curved building forms a perfect horseshoe — toeclip, caulks, and all.

evidence suggests that it was built in the 1920s, perhaps as a hunting stable. It is certainly admirably adapted to its present function. The Tartan-floored stalls are airy and measure twelve by twelve feet at the rear, losing about sixteen inches on the face; their doors are centered. The courtyard, behind its high stone wall, faces southeast, catching the sun, yet protected from the worst assaults of heat or wind. These stalls have sheltered generations of Mr. Mellon's high-bred yearlings, including Fort Marcy, Arts and Letters, and the incomparable Mill Reef. Mr. Mellon also has an enclosed circular exercise barn, where foals and their dams can be safely turned out in severe weather.

The long stone stable at Xalapa, in the Bluegrass country, its North Italianate colonnade following the outer side of the curve, is another variation on the circular theme. The arc is flattened, so that the building forms a segment of an enormous imaginary round. The outward-facing stalls are reminiscent of the scheme of the vanished stallion barn at Dreamwold, in Egypt, Massachusetts, which was part of a Queen Anne farm-village complex. The object here was that no horse should see his neighbor. This unquestionably inventive arrangement was assessed in the *Architectural Review* of September 1902 as "of little consequence, as [Mr. Lawson's] stallions, though high spirited, are neither morose nor ferocious."

127

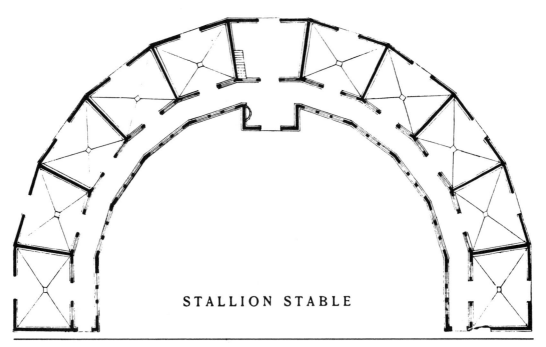

STALLION STABLE

The plan of the stallion barn at Dreamwold, Egypt, Massachusetts, ca. 1902.

Dreamwold, as ambitious and in terms of buildings as extensive as the Hartman or Anna Dean empires, was the creation of Thomas W. Lawson, the Copper King, who played a major role in the development of America's nonferrous metals industry, moving and shaking the world of finance. He amassed a tidy fortune. He was a yachtsman, several times involved in the America's Cup races, of which he co-authored a history, and was deeply interested in the breeding of domestic animals.

The Boston firm of Coolidge & Carlson were the architects of Dreamwold, but the color and style of the buildings were set by the owner. They were intended to harmonize with the seventeenth- and early eighteenth-century houses of the neighborhood. Moreover, no detail of animal housing or comfort or of farm economy was too small for Mr. Lawson's attention. "Everything," he declared, "must be heavy, strong, simple, and quiet; if four by six will do, make it six by eight." Every building, from the elegant family "farmhouse" to the dovecotes, was shingled in weathered gray with green shutters and white trim and carried a broad shallow gambrel roof. There were two sets of kennels for the two breeds raised on the place; a huge cowbarn-dairy and subordinate buildings; poultry houses, utility buildings, and, of course, at the heart of the farm, the stables. The farm crest was Pegasus, bridled and held by a man of heroic cast, its motto, "Beauty, Strength, and Speed."

Draft horses, carriage horses, and racehorses were raised at Dreamwold on a lavish scale. There was a driving park, containing a nine-acre polo field, surrounded by a training track, with a racetrack on the perimeter. In addition, there was an indoor school, measuring one hundred and eight by one hundred and seventy feet. The racing stable was eight hundred feet long, the draft horse barn almost as large, and the broodmare and foaling stables were of appropriate size.

Dreamwold as an entity did not long outlive its builder; although a number of its characteristic structures survive and are in use, the stallion barn is unfortunately not among them. We know, however, that the stalls were fourteen by eighteen feet, with solid dividers and front grilles, floored in two-inch spruce laid over tarpaper on seven-eighths–inch spruce, which was in its turn supported by hemlock planks set in six inches of tar concrete over two feet of crushed stone. The top layer of flooring had to be replaced every six months to two years, but this was held to be a negligible disadvantage compared with the perceived drawbacks of earth, clay, macadam, concrete, or brick flooring. We know also, at least by inference, that an army must have been involved in the original raising of Dreamwold. The entire complex was conceived, built, and inhabited in a little more than a year.

There are a number of Thoroughbred training barns built on an oval or given curved ends, so as to better accommodate the tanbark or sand track encircling the central double row of loose-boxes. Some of these training stables are more or less elegant versions of the Gallatin, Tennessee, barn described on page 106, but the track is often enclosed, especially if there is a loft overhead requiring an outside bearing wall. The indoor track thus provided is perfectly satisfactory for jogging exercise even if it describes an oblong, which it generally does, but an oval configuration means that actual racetrack conditions may be pretty exactly simulated.

One of the most attractive examples of a curved-track training barn is at Clovelly, the Robin Scully farm near Lexington, Kentucky. It was designed in

A view of the covered schooling track at the Clovelly barn.

1928 for Joseph E. Widener by Horace Trumbauer (1869–1938), who was responsible for a great many civic, commercial, and residential works for the Widener interests, in Pennsylvania and elsewhere. Trumbauer was self-trained and worked brilliantly in a variety of styles; perhaps his most extraordinary achievement is the original campus at Duke University, which constitutes, particularly in its chapel, the culmination of twentieth-century collegiate Gothic Revival.

Nothing could be further removed in concept or aspiration from that medieval exercise than the deceptively modest Clovelly stable. The building is of frame construction. The enclosed perimeter, lined with windows, is an oval tanbark track measuring one eighth of a mile; a simple tray ceiling gives extra height and a pleasant interior perspective. Two rows of sixteen stalls each, plus a central feed room, face one another across the open space inside the oval. Their floors are clay, and they open on both yard and track. They are fitted with solid doors surmounted by transoms and have slatted inner doors for use in warm weather. Roomy lofts with several dormered access doors overhang the stall ranges, and there is a hay drop in every stall. A small office, glassed all around for observation, stands in the center of the court. The oval is pierced at both ends. On the approach side, the doorway fits beneath the eaves, but at the farther end is a high access under a raised penthouse which allows free passage for trucks and hay wagons. Where these entryways cross the tan, a floor of ten-by-ten poplar sills, laid perpendicular to the track, ensures against undue wear.

There is virtually no applied ornament, save a pleasant contrast in the paintwork, to be seen in this composition, which relies on the skillful and unob-

The training barn at Sagamore Farm, Glyndon, Maryland.

130

trusive juxtaposition of geometric forms for its success. Perhaps, in Mr. Widener's time, Stagehand, the first three-year-old to win the Santa Anita Handicap, or the great steeplechaser Bushranger occupied stalls here when they were learning to accept a rider; certainly many good horses have been around the well-graded oval since.

The Clovelly track measures a furlong. The equivalent covered oval at Alfred G. Vanderbilt's Sagamore Farm in Maryland measures two — a full quarter mile. Ninety horses can be stabled there, either in the curved range lining half the oval, or in the straight blocks attached to the far perimeter. All these stalls are twelve feet square, wood-floored, and open onto an interior aisle rather than a shedrow. There are four entries to the interior yard, one at each end, and one through each of the stall ranges, so that no horse is ever a great distance from some point on the track, while feed and manure vehicles can get in or out conveniently. The building was put up in 1930, one of many stabling facilities on the property. These include an additional ninety stalls. When Mr.

Interior of the Sagamore training barn, in a painting by Vaughn Flannery. Private Collection.

Hermitage Farm, Oldham County, Kentucky. The long, plain inside barns of this Thoroughbred nursery, somberly creosoted to match the surrounding fences, span the farm road like covered bridges, giving direct vehicular access to every stall.

Vanderbilt's racing string numbered more than fifty, as it did for many years, the training barn stalls saw heavy use; it is less busy today, as the owner's other interests take too much of his time to keep so many horses in training. Nevertheless, the track itself is still an invaluable part of the operation, and the blood of Discovery, Native Dancer, and Restless Native flows in the veins of the colts and fillies that work out there.

Mr. Vanderbilt's contributions to the Turf are staggering. Like many another sportsman, he has labored long and willingly on behalf of his favorite avocation. The giant Sagamore track and stable is on a scale appropriate to its owner's stature in the history of Thoroughbred racing.

The long airy barn at Will Rogers State Park and Monument, on the Pacific coast near Los Angeles, is a far cry from the great nursery installations of East or West, yet the tall round chamber that divides the building in two recalls, in a gentle way, the vainglorious facade of Savage's International Stock Food Farm (page 107). Rogers moved the old barn building from west Los Angeles to its present site in the 1920s and inserted the rotunda to give himself a private spot to practice trick roping. It may be closed off from the stables proper and the outside world by doors that follow the curve of the walls. The stable is now chiefly tenanted by local boarders, and the ring is used as a longeing arena. However, we are told that Rogers himself never used it for its intended pur-

132

pose, but practiced his roping wherever the spirit moved him, even in the house, where, under specially raised ceilings, he snaked his lariat among domestic hazards without stirring anything except, perhaps, some wifely apprehensions.

In the builder's day, the Rogers stable housed family pleasure horses and polo ponies. It is a clever adaptation of a conventional cool-climate scheme to local conditions. A center aisle runs the length of the whole building to end openings. The clay-floored stalls open onto the aisle, and windows line the ranges from wainscot height. There is no loft or ceiling; feed is stored in the barn, hay and bedding in a separate building. Under the eaves is an uninterrupted strip of louvers; with doors and windows open, the barn becomes, in effect, no more than a set of pens sheltered only from the sun. The roof of the circular chamber is supported by a wagon-wheel truss system, providing an appropriate ambience for the rehearsals of the great cowboy humorist. Perhaps it was this eminently suitable quality that caused him to eschew it.

Polo is still played at the Park on Thursdays and Saturdays, and the public is welcome to attend. Rogers was a true lover of the game; he came to it with all the enthusiasm of the adult who discovers a hitherto unsuspected joy. Watching an exciting chukker at the Park, one remembers his pleasure and the wild tales of his daring on the field.

Left: Stable at the Will Rogers State Park and Monument, Los Angeles County.
Right: Interior of the circular chamber at the Will Rogers stable.

133

The adobe stable at Lazy Acres Ranch, Tucson, Arizona.

Wide variations of temperature take place rapidly in the dry air of the Southwest; proper ventilation is of vital importance in stable buildings, but some protection from chill must also be provided. The Rogers stable, in a heavily irrigated area, provides one satisfactory answer to the problem, but the adobe barn at the erstwhile Lazy Acres Ranch outside of Tucson, Arizona, exemplifies a solution, well suited to desert conditions, that harks back to the aboriginal owners of the land. Lazy Acres was established in the 1930s by Melville Haskell, a towering figure in the early annals of the modern Quarter Horse. All the buildings on the place were built of adobe, and constructed by traditional methods. Bricks and mortar are of the same substance, and wood was employed only for roof supports and millwork trim. The stable building is a simple oblong with a center aisle, positioned to catch the prevailing winds. The floor is tamped clay, and the lower stall-walls are a part of the fabric. Wooden grillwork surmounts the adobe partitions, but this may not be original. Within the building, the thick walls and heavy roof keep out the heat of the day and turn back the chill after sundown, while the soft color and severe lines of the exterior chime with the surrounding landscape.

When the stable was photographed for this book it was under lease to a hunter-jumper trainer and instructor, Judy Trotter Martin, who describes it as one of the pleasantest barns to work in she has encountered. She is, unfortunately, the last to enjoy it. The present owners have scheduled the demolition of all the existing buildings to make way for a complex of condominiums. The loss is a sad one, for few such faithful re-creations of this ancient building system survive.

134

Hunting stable at Knole, Old Westbury, built in 1903 for Herman B. Duryea by Carrère and Hastings. The arch piercing the building is the only free access to the court, onto which stalls, utility rooms, and grooms' quarters open.

Attractive and agreeable as the Rogers and Haskell stables are, they were not fashionable. According to *Sporting Stables and Kennels*, the "inside" stable barn with a row of stalls to either side of a single aisle is "very economical and easy to run," but "not very attractive for a private stable." Of twenty-six stable plans illustrated and discussed in this volume, the authors show only nine that do not at least partially enclose an interior court, and of these one is the horseshoe noted earlier and two are conventional Thoroughbred training barns. The quadrangle was by long odds the preferred scheme, East and West.

The Riding Club and Stables at Hope Ranch, that venerable residential enclave near Santa Barbara, combines the advantages of the "economical" and the "attractive." It dates from 1930 and is the work of Reginald D. Johnson (1882–1952), a transplanted New Yorker particularly remembered for his residential designs in the Santa Barbara area. He successfully blended a Mediterranean ambience with traditional Anglo-Saxon forms, and the Hope Ranch stable is a striking display of this talent. The entrance front, a freestanding plastered building with low-pitched tile roofs and a cupola, is pierced by a two-story Moorish arch. Behind this frontispiece is a fenced riding ring, surrounded by plain, self-contained board-and-batten "inside" barns which add the attractions of a closed yard to a simple and practical stall scheme. The Riding Club is an active operation and is the home of a group of handsome pleasure horses of various breeds.

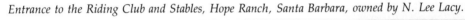

Entrance to the Riding Club and Stables, Hope Ranch, Santa Barbara, owned by N. Lee Lacy.

Los Altos Hunt stable, Paicines, California.

An excellent illustration of the half-court stable is the headquarters of the Los Altos Hunt, which was once the stallion barn on the Robert Law Ranch. It is now in the Almadén vineyards; foxes, hounds, and field alike tread (or thread) the vines in hunting season. The stableyard, beneath the spreading shade of a great boxed-in oak growing in the center, is swept sand. The unusual louvered stall doors opening onto the yard have side-hinged transom panels, so that considerable variation in temperature and ventilation may be obtained. The white, plastered building, incorporating feed room, tack room, office, and so forth in the wings, is as simple and functional as one could look for; the flower stands between the stall doors, and the Art Deco crestings at the eaveline, add a pleasantly insouciant air to an otherwise somewhat austere composition.

137

A larger and more ambitious hunt stable, and another variation on the quadrangular plan, is the home of the Mill Creek Hunt in Illinois. The complex includes stabling, the Kennel-Huntsman's residence, and a pleasant clubhouse on axis with the entrance, all disposed on the perimeter of an open gathering place; the kennels are behind the clubhouse. The stall ranges at the corners are curved, rounding off the enclosure; tack and feed rooms are located next to the stall groups for easy access. The frame buildings are unassuming, laid out for use rather than for show, but the total effect of the well-planned composition is charming as well as practical.

The 1928 red-brick stables at Pebble Hill, Thomas County, Georgia, were originally cattle barns, designed by Cleveland architect Abraham Garfield to house a prize Jersey herd and a self-contained dairy operation. Garfield was commissioned by the owners, Mr. and Mrs. Perry William Harvey, to blend the flavor of the Lawn at the University of Virginia with that of a Normandy village. The result, if a tour de force, is a triumph.

The shaded grassy island between the buildings was outlined by a serpentine wall. Cottages, apartments, feed storage, byres, and work spaces for the dairy were integrated into the scheme by the skillful juxtaposition of Jeffersonian pavilions with silo towers, dormers, and turrets. Mrs. Parker Poe (1897–1978), Mrs. Harvey's daughter, revamped the complex in the 1940s. Clay-

Members of the field leaving the Mill Creek Hunt stables, Wadsworth, Illinois, on a crisp October morning.

138

Central pavilion at the Pebble Hill stables, Thomasville, Georgia.

The brick stable complex at Pebble Hill.

floored box stalls, some opening onto an inside aisle and others onto the court, replaced the rows of stanchions; a carriage room and stable utility rooms took the place of the milk-processing areas. So successful is the adaptation that one is surprised to learn that the barns were erected for quite another purpose.

Mrs. Poe was a sportswoman. A crack shot and a foxhunter (she followed her own pack of Walker hounds in both Kentucky and Georgia), she was in her youth a polo player and a jump rider of international caliber. Distinguished racing stock came from Shawnee Farm, her Bluegrass Thoroughbred nursery, and in later years she developed an interest in Appaloosas and Quarter Horses, which she bred at Pebble Hill. At the time of her death the stable area was policed at night by a particularly fine-headed Appaloosa stud. His arrogant carriage and dramatic way of going (wholly belying his gentle, friendly nature) would unquestionably have intimidated any intruder.

Now maintained by the Pebble Hill Foundation, the stable is unoccupied, but when plans for the future are implemented and the plantation is open to the public, perhaps visitors will again be greeted by inquisitive equine heads framed by their half doors. The picture will be incomplete without them.

The Rolling Rock Hunt stables in Ligonier, Pennsylvania, are probably the most imposing buildings of their kind and period in current use. Despite certain similarities in configuration, they are totally different in impact from the stables at Pebble Hill. Laid up in dark gray stone and roofed in slate, they form three sides of a hollow square, which is completed by the gateway wall. The exterior of the square is skillfully interrupted by gabled projections and the high roof and cupola of the tall service block, so that it appears to be several adjoining structures rather than a continuous whole. Entrance to the graveled yard is gained through a wicket in the eastern stall range or through the main gates, which are hung on square piers topped with guardian owls. All the twelve-foot-square stalls, of which there are twenty-four, open to the court and to an interior aisle, except for three shedrow stalls on the east, which give onto the entrance drive only. Hay, feed, and bedding are stored in the service block, which also holds tack and blanket rooms, offices, equipment storage, and the like. An open gallery runs through the service block, and on each side of it, where the ground falls away, are traps through which stable muck may be dropped into waiting carts.

The architect of Roling Rock was Benno I. Janssen of Pittsburgh. He took pains to make the interior of the stable as fine as the exterior and left no detail unattended to, from stall hardware to drying arrangements. The wide aisle is flagged, vaulted, and splendidly lit by big windows. The stall floors are clay, over a layer of crushed stone and another of charcoal; the grillwork is handsome, and tie rings, rails, and racks are all decorative and well-situated.

The stable is still in use. It is a thoughtful layout, although housekeeping chores relating to the building itself now present a considerable challenge. Nevertheless, although gleaming brass, polished wood, scrubbed stonework,

140

COLOR PLATE 16. *A horse barn at Hartman Farm, Franklin County, Ohio.*

COLOR PLATE 17. *The Long Barn, Kentucky Horse Park, Lexington.*

COLOR PLATE 18. *Abandoned Gilded Age stable at The Reefs, Newport, gutted by fire and vandalism, but now under the protection of the Rhode Island State Department of Natural Resources.*

*Stall range, La Bagatelle, Old Westbury, New York, from the entrance gate (above) and from the doorway of the house (*COLOR PLATE *19).*

COLOR PLATE 20. *The Amasa S. Mather stable, Chagrin Falls, Ohio. The stylized Norman building was designed in 1919 by Harrie T. Lindeberg.*

Color plate 21. "The Stable," Houston. Only the configuration of the doors, high enough to admit horse and rider, reminds the visitor that this elegant party house was once a working carriage house and stable. It was designed in 1930 for Mr. and Mrs. Harry C. Wiess by John F. Staub, who was also responsible for the 1955 conversion.

COLOR PLATE 22. *Semicircular yearling barn, Sundridge, Upperville, Virginia.*

*The polo stable at Caumsett,
Lloyd's Neck, New York,
during the early 1970s (above)
and after restoration
(COLOR PLATE 23).*

COLOR PLATE 24. *The 1922 Rolling Rock Hunt stables, Ligonier, Pennsylvania, built by General Richard K. Mellon, M.F.H. Exterior stalls.*

The courtyard at Rolling Rock.

and swept gravel mean little to the animals taking the sun over their half doors, they lift the spirit of the human visitor. Plan xxvi in *Sporting Stables and Kennels* shows a partial view of Rolling Rock. It has pride of place at the close of the stable section, and is accompanied by two photographs and a paragraph of unstinting praise. With becoming modesty, the authors have described Vernon Manor — the building which led to their collaboration — in Plan xxiv, leaving the Marshall Field polo stable at Caumsett, on Long Island (discussed below), and Rolling Rock to climax their dissertation.

Had it not been for the happy association between architect and client developed during the design and construction of Vernon Manor, Mr. Gambrill's

141

Peapack, New Jersey, estate, we would not now have the pleasure of consulting their truly valuable compilation. The book may arouse ironic mirth as well as (covetous) nostalgia in the modern reader, especially in regard to domestic arrangements for staff, but it contains a great deal of still sound advice.

Richard V. N. Gambrill was a foxhunting man, pursuing sport on both sides of the Atlantic, and was Master of a Beagle pack (eventually the Vernon and Somerset Beagles) for forty years. He drove the famous Vanderbilt grays to his road coach *Defiance,* and was not above putting one of his hunters to trap or sleigh. The grays were succeeded by a team of Hackneys, and these in turn by four smart cobs; *Defiance* was a well-known sight at shows and hunt meetings.

The family shared his interest, so that the Vernon Manor stable was a center of activity. It is highly improbable that anyone would build a domestic stable on such a scale nowadays, but then a modern installation is unlikely to require a carriage room and a sleigh room, together with harness room and vehicle washroom, in addition to stabling for twelve hunters, their appurtenances, and four coach horses. A pleasant center-hall Colonial stud groom's "cottage" is connected to the rear of the building by a covered way under the eaves of the farrier's shop; "strappers," or undergrooms, lived above the heated tack and washrack area. The arrangements show an astute blend of practical and aesthetic value.

Mr. R. V. N. Gambrill, M.B., on the box of the road coach Defiance *in the stableyard at Vernon Manor, in a painting by F. B. Voss.*

Detail of the facade of the Vernon Manor carriage house.

Venture, Viking, Vogue, and Vanity in their stalls at Vernon Manor.

The building turns its back on the north wind, catching the sun in the yard. Each stall, floored in cork brick, has a half door, a sizable window opening under the arched arcade, and a small service door, supplemented by a barred window frame, which gives access from the corridor behind it. The stall partitions are solid to within two feet of the ten-foot ceiling; each stall has its own hay drop, which rises into the loft and may be closed with a hatch. Running water was supplied to every stall. The feed room is at one angle of the service passage and the drop to the manure cart at the other; the stall area is confined to the inner half of the yard. The tack cleaning room, washrack, tack room, harness room, and sleigh room occupy the outer half of the east wing; opposite them are the carriage house, vehicle wash, and closet room. The carriage house measures thirty-six by sixty-three feet (Mr. Gambrill, in addition to the red and black *Defiance*, had a remarkable collection of other vehicles, many of them inherited) and extends outside the west wall for more than half its length, thus preserving the architectural balance within the courtyard.

143

In style, the stable is an understated reflection of the formal Georgian manor house, and in spite of an undoubted elegance, has a warm and welcoming air behind its low brick wall. Vernon Manor has changed hands, but the future of the stable seems assured. It is so well arranged and inviting that urgent schemes for preserving it and restoring at least some part of it to active life spring vividly to mind.

The Marshall Field stable at Caumsett, which for some years stood virtually abandoned, suffering not only from casual vandalism but from the fortunately minor effects of several fires set in an upper room as part of a local fire department teaching program, has been rescued. Now within the confines of a state park, it is an equitation center emphasizing combined training, operated under lease from the State Parks Department of Long Island by Mr. and Mrs. John Russo. Mr. Russo is gradually restoring the building to its former state, doing much of the work himself. Happily, its fireproof construction prevented damage to the sophisticated and complex system of ductwork for heat and ventilation. Most of the elegant stall fittings were preserved, because they were wisely painted black or removed to storage when the stable was originally vacated. Much has been done, and there is more to do, but the essential structure is sound, and as satisfactory as ever.

It was designed in the 1920s by John Russell Pope (1874–1937). The roster of outstanding public and domestic works by this distinguished architect is as-

Interior of the Caumsett stable, with air scoops in hot-weather position. See also color plate 23.

tonishing: among the highlights of his career are the National Gallery of Art, the Jefferson Memorial, and the Payne Whitney Gymnasium, last of the Gothic buildings at Yale. The entrance courtyard at the Polo Stable, fronted by a brick wall with iron gates, is flanked by flanged wings containing offices, tack and feed rooms, blanket and drying rooms, sanitary facilities, and two comfortable cottages. The central block, or stable proper, rises a full two stories over the stalls. Arched openings to front and rear frame the view from the court to the broad expanse of the old polo fields, and there are additional wide doors at either end of the aisle.

The big box stalls, of which there are sixteen, including a double wash stall, are floored, like the aisle and corridors, in brick, and are sloped to central drains. These stall drains are connected to cleanout valves at the end of each run, which allow them to be flushed out periodically with a high-pressure jet of water. The stall posts, each with its brass tie ring, are capped with heavy brass balls; the door latches are heavy brass-finished mortise-bolt fasteners, with recessed thumb slides. All the interior fittings, from drains to bridle brackets, were supplied by the James W. Fiske Iron Works of New York.

Hay and bedding are stored in the mezzanine lofts, as is feed. There are drops to the feed room and at either end of the stall range, all of which are concrete shafts which may be sealed off with metal fire doors. There is, in fact, very little wood used in the building, yet it is an inviting place: lofty ceiling, ocular windows, and all. The duct system is quite extraordinarily effective. In summer the temperature may be maintained at a point at least fifteen degrees lower than the outside air. In winter, although the system that heats the rest of the building can also serve the stable area, it is never necessary to turn it on if the stalls are fully occupied, in spite of the high ceiling.

As subsidiary domestic buildings, stables on the order of Vernon Manor, Woodlea, and Caumsett have outlasted their time; no beauty of proportion, luxury of interior finish, or excellence in construction can perfectly preserve them in the face of the maintenance problems of today. The true survivors are the comparatively modest home stables built to house a handful of horses in the 1920s and 1930s, which rely on unemphatic vernacular interpretations of more sophisticated styles to give architectural character to their strongly pragmatic designs. The owner-designed 1938 residence and stable at Southlands Farm, in Dutchess County, New York, is modeled after an island fisherman's home in which the owner sheltered after being shipwrecked in the Baltic Sea. One of the few schemes of that period to bring stable and house under the same roof, it conveys, from the entrance front, a truly European sense of walled privacy. In the front rooms one is unaware of the proximity of the stable, but the long, well-lit living room is opposite the shedrow, so that people and horses can take an interest in one another across the courtyard. The commodious foaling stall, with its louvered corner, may be converted into three straight stalls at need, and there is ample hay storage overhead.

Rear elevation of "The Barn," Southlands Farm, Rhinebeck, New York, in a watercolor by Olin Dows.

A private hunting stable in the Radnor country of Pennsylvania, built in the 1930s to the owner's design.

The Ambrose F. Clark polo stable, Aiken, South Carolina. The exterior has been somewhat altered.

The courtyard stable at the A. E. Reubens' Hasty House Farm, Toledo, Ohio. Mrs. Reuben's show hunters, including the well-remembered Darlington, were housed here. The racing string trained elsewhere, but photographs of Hasty Road, Oil Capitol, and other stakes winners share the tack room walls with a brave array of horse show ribbons.

Stable in the old Westchester Hunt country, from the garden. Opposite: interior of the stable block.

The stable wing of the pleasant frame complex of outbuildings in the accompanying photograph is also still tenanted. This grouping supports a home in the country of the old Westchester Hunt, which flourished on the New York–Connecticut border before the war. The central block is the garage. A porte cochere connects it to a utility barn on one side; on the other, a breezeway divides a cottage attached to the garage from the stall area. The tack room is part of the cottage, giving it the advantage of tempering heat and hot water; the breezeway doubles as a washrack. The stalls are designed to open onto driveway and aisle; only the inside doors are in use, but the outer doors remain available in case of emergency.

148

There are six stalls with a loft over them, supplemented by additional storage over the utility building. In the aisle, the ceiling peaks to the roof, giving a charming perspective and improving ventilation. On the garden side the walls are glassed like those in a training barn; when they are open, the aisle becomes, in effect, a shedrow. The whole arrangement is neat and practical. The horses are close to house and cottage, but far enough removed for comfort. All estate facilities are concentrated in a compact area, and the farm group effectively screens the house from the driveway. The composition has a gracious air, yet, like "The Barn" (page 146), it exemplifies Alexander Pope's definition of simplicity: "the Mean between Ostentation and Rusticity."

The stable at Pook's Hill, as illustrated by Pierre Brissaud on the cover of House & Garden, *August 1939.*

The stable at Pook's Hill, residence of Mott B. Schmidt, A.I.A., designed by the owner.

VIII Themes and Variations

*There is nothing so good for the inside of a man as
the outside of a horse.* —ANONYMOUS

The above quotation has been variously attributed; it seems best to give credit
to the coiner of most of our pithier sayings. Certainly the aphorism has never
been more heartily accepted. We all live in horse country now; the revival of
mounted police troops has even brought the creak of saddle leather back to the
darkest urban canyons. This phenomenal equine resurgence started after the
Second World War. Its extent is most readily substantiated by the enormous
increase in registered horses of all breeds and by the appearance of hitherto
unfamiliar breed names; the end is not in sight.

The Arabian is a case in point. The Arabian horse was imported into Europe for centuries before the Americas were colonized. Virtually all the modern
light-legged breeds have Arabian ancestry. The "Drinker of the Wind," the
"Skimmer of the Sands," has always had a legendary quality, yet in spite of
beauty, stamina, and remarkable constitution, the Arab was for many years
honored more in this country for its contribution to other breeds than for its
own singular and admirable characteristics.

Nevertheless there have always been those to whom the "perfect" Arabian
was the "perfect" horse. At first there were only a handful of Arabian breeders
of any consequence: Spencer Borden, J. A. P. Ramsdell, Randolph Huntington,
Albert W. Harris, William Robinson Brown, and General J. M. Dickinson, plus,
of course, Homer Davenport and Peter B. Bradley, who backed Davenport's
Roosevelt-sponsored 1906 shopping expedition to the desert and was his associ-
ate in the Hingham Stock Farm. Arabians were raised from New Hampshire to
California, but in 1925, when William K. Kellogg established his Pomona ranch
with the intention of improving and promoting the breed, the Arabian Horse
Club of America listed just five hundred animals in this country, living or dead.
Nowadays that figure does not even approach the number of new entries in the
Registry every year, and there are serious breeders in every one of the contig-

151

uous states. A considerable degree of credit for this must go to the Pomona installation and its founder.

The Kellogg Foundation has established or supported many valuable projects; extensive programs for handicapped children, bird sanctuaries, and model farms owe their prosperity to the persistent popularity of dry cereal breakfast foods and Mr. Kellogg's eleemosynary propensities. The love of Arabian horses that inspired his California venture stemmed from a childhood affection for an unpedigreed paint pony, all of whose better features were attributed to Arabian blood. It was a disappointment to Kellogg to learn that skewbalds, while historically possible, are about as common as hen's teeth in the Arab registries.

Having acquired the ranch, he stocked it in wholesale fashion, buying heavily from W. R. Brown's Maynesboro Stud in New Hampshire, from the Hingham Stock Farm, and from Lady Anne Blunt's Crabbet Park Stud in England; the first Skowronek blood in this country came to Pomona. Among the sires bred or acquired by the Kellogg Ranch, whose names recur in noted modern pedigrees, are *Raseyn and his son Ferseyn, Abu Farwa, and Khaled. On the advice of Harris and Brown, the Stud emphasized conformation and performance rather than consistent ancestry, and trained all the animals to show, in one way or another, the versatility of the breed. Visitors are still welcomed every Sunday afternoon to watch Arabians jumping fences, doing "liberty" circus acts, performing the Spanish Walk, the rack, or the slow gait, going smartly in harness, and demonstrating the skills necessary to a cutting horse.

In 1932 the ranch was donated to the University of California and became the Kellogg Institute of Animal Husbandry. In 1943 it was transferred to the War Department as the Pomona Quartermaster Depot (Remount). In 1946 the Remount Service passed to the Department of Agriculture, but when the Remount operation was wound up, the Department commenced to disperse the herd. The Kellogg Foundation promptly repurchased the entire facility, and in 1949 transferred it to the State of California, specifically to the California State Polytechnic College (now University) on a renewable lease, subject only to the twin conditions that the Stud be maintained to a high standard and the Sunday shows continued. Eventually the imposing 1926 stable (page 209), the first building erected on the ranch, was absorbed into the general Cal Poly campus, and the Foundation underwrote the present handsome facility, which was completed in 1974 to the designs of the California firm of Neptune and Thomas. It conforms in general plan to the earlier building, at the same time presenting an impressive contemporary facade. The architectural interpretation of the familiar quadrangle is extremely effective. The overhead service monorail of the original stable has not been reproduced; there is ample aisle and alley space in the new facility for motorized service, and the feed and tack rooms are conveniently situated. The narrow pipe-fence pens in the courtyard are interesting. Set over drain pans, they are just large enough to restrain a horse. They facilitate minor

Interior court of the 1975 stable at the California State Polytechnic University, Pomona.

veterinary treatment and the examination and artificial insemination of mares, and make convenient washracks as well.

Cal Poly faculty and students share in the breeding and schooling programs. Through the courtesy of private owners, mares are selectively bred to outside stallions, but the Crabbet-Maynesboro lines have not been lost. Because the herd has been under state or federal management since 1932, Kellogg Arabians have rarely, as a matter of policy, been shown in competition with privately owned animals, but the influence of the aggressive and intelligent breeding program pursued at the ranch is still felt.

When Edward Tweed saw his first Arabians, he felt he had "seen the essence of what a horse should be." He devoted more than a quarter of a century to developing and improving the Arabian horse in America; he was the founder of the renowned Scottsdale Arabian Show. He purchased his foundation sire, Skorage, in 1950 from Daniel C. Gainey, then doyen of the Arabian Horse Club Registry of America, and established the young stallion, with the nucleus of a broodmare band, at a ranch near Scottsdale, which he called Brusally after his son and daughter.

Skorage achieved a remarkable show record, and his get measured up to his high standard. When asked the secret of a successful breeding program, Mr. Tweed might have paraphrased Alfred G. Vanderbilt's tribute to Discovery and replied, "Breed a mare to Skorage." In the 1960s Mr. Tweed began importing Polish stock, with a view to increasing size and bone in his herd. Polish Arabian studbooks have been kept since the eighteenth century; the great Raffles, a name almost synonymous with Crabbet bloodlines, was sired by the Polish-bred Skowronek. Brusally has since been distinguished by racing, performance, and halter champions in which Polish blood has predominated, although the Anglo and Egyptian strains have not been lost.

All the buildings at Brusally are built in a pleasant Mission idiom to the design of the owner, whose early career was in architecture and engineering. The thick white-plastered walls, low-pitched tile roofs, and exposed heavy beams in the stable buildings recall the cool comfort of the Lazy Acres adobe (page 134). They are rectangular inside barns with spacious aisles and conveniently arranged utility spaces. Wagon-wheel fixtures, like those that illuminate the living room of the residence, hang in the aisles, so that in spite of their simplicity the stables have an air of being an extension of the home, appropriate to the companionable relationship between horses and people at Brusally. Miss Shelley Groom manages the operation for her grandfather.

The oldest stables at Brusally Ranch, Scottsdale, Arizona.

154

Scottsdale is a hub of the Arabian horse industry in the United States, and what is perhaps the largest and most ambitious complex devoted to the breed is located there. Lasma Arabians, which breeds almost exclusively from Polish stock, is the property of the La Croix family. The establishment consists of stabling for more than one hundred and forty horses, with appropriate service areas: a "controlled-environment" equine clinic with operating and recovery rooms, convalescent stalls, laboratory, and elaborate X-ray equipment; examination and breeding facilities (all breeding is by artificial insemination); extensive pasturage, including eighty private paddocks, served by an underground irrigation system; several training areas of differing types where horses may be not only schooled but exposed to the sort of distractions they will encounter in the show ring; and a vast sales building (again with a controlled temperature system), which contains auction auditorium, reception lobby, telephone room, grooms' or owners' apartments, dressing rooms, tack rooms, grooming and wash stalls, veterinarian's office, and a roomy sales barn.

Dr. Eugene E. La Croix, a practicing surgeon before coming to devote all his time to Lasma, has designed the interior of the barns to meet an impeccable standard of cleanliness, so that they rather lack the cozy quality generally associated with stables. Still, there is no lack of provision for horse comfort; every resource of modern technology is invoked for their health and protection. The stable area is attractively planted, and the low-pitched buildings and varied schooling areas form a pleasantly understated composition which belies the magnitude of the serious business they represent.

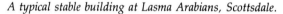

A typical stable building at Lasma Arabians, Scottsdale.

The Polish-bred *Bask was the foundation sire of Lasma Arabian horses as they are today, and the foundation as well of what has become a deeply satisfying and profitable venture for the La Croix family, every member of which is involved in some facet of the organization. *Bask, like many Arabians hale and hearty into his twenties, presided over Lasma, viewing his get and grandget with an alert and patriarchal eye.

Anyone planning to invest in a horse must look for health and soundness, and will therefore carefully observe the prospect and probably have a favorable opinion confirmed by a veterinarian. Temperament and performance record, if any, will also be weighed; for most purposes, handsome is as handsome does. In the Arabian, however, star quality has always been required. The immediate and fundamental visual appeal of the animal is very nearly as important as the results of subsequent investigation. The auctions held at Lasma Sales Center are as carefully staged as a Broadway extravaganza. The value of this orchestration, expressed in the prices fetched, has long been understood. When James C. Lewis, having been caught by Arabian charm on his visits to Scottsdale, established a breeding operation on his farm in Albemarle County, Virginia, he carefully designed his stables with the importance of initial impact in mind.

The stallion barns at Lewisfield, each stall opening into a paddock, were set at the lower end of the property. The great *Nizzam and, later, *Touch of Magic had individual stables in special double-fenced paddocks, where they could be seen grazing peacefully or breasting the slope of their steep enclosures at a powerful gallop. The accommodations for broodmares, weanlings, and show stock were set at the top of the long drive that winds through rolling pastures to the crest of the hill. (In one of these paddocks, with comfortable run-in shed and ample water, the versatile Alyfar, bred at the Kellogg Arabian Ranch and brought to fame by Arthur Godfrey, lived happily to a remarkably advanced old age.) From the gravel sweep before the stable entrance, one looks out across the valley to the ever-changing prospect of the Blue Ridge; a horse turned out in the exercise ring by the forecourt is displayed against a backdrop no human art could improve upon.

The stable complex itself neatly combines practicality and showmanship. Two plain, lofty, inside barns are separated by a gracious office and reception center backed by a covered arena, which may be observed through a vast plate-glass window in the elegant and comfortable lounge. The central block, fronted by a fine portico taken from a derelict antebellum mansion, has high ceilings and terrazzo floors, a convenient kitchen, and a projection screen that may be drawn down over the observation window. During the Lewis ownership, the gallery was furnished with leather chairs, and with massive old pieces in dark wood which supported some part of the builder's collection of bronzes; the walls were hung with portraits of distinguished Arabians of the past. When the curtains were drawn back from the sweep of glass, the spectator faced a broad cabinet on the far wall of the ring, filled with rows of ribbons brought home by

156

Stables at Lewisfield, Albemarle County, in the early 1970s.

Lewisfield produce. Like the barns, the arena is a rectangular building, painted a warm brown; within, it is lined by a sloped and curved wainscot, which offers a passage on the inner wall giving access to the gallery.

Each barn may be operated independently. Hay storage is over the clay-floored stalls; the unusually wide aisles are open to the roofs, which support cap ventilators, and in fine weather are also open at both ends. The big foaling stalls are in a separate building. The Lewisfield horses are dispersed, and the farm has changed hands, but the blood of *Nizzam and of the silver-coated *Touch of Magic still flows in the veins of champions.

The healthy climate, rich pastures, and rolling terrain of northern Virginia promote the natural development of muscle and wind in young stock and help to keep broodmares in trim. The area produces outstanding horses and ponies of many breeds, but since the early nineteenth century these rounded hills have been first and foremost a cradle for Thoroughbreds. A relative newcomer to the list of famous Virginia nurseries is Catoctin Stud, named by its owners, Mr. and Mrs. Bertram Firestone, for a creek that waters the property (see title page). It

157

The broodmare barn at Catoctin Stud, Waterford, Virginia. See also title page.

158

was established in the 1970s. Mr. Firestone, who is also the proprietor of Gilltown Stud in Ireland, began his American Thoroughbred operation rather modestly, but soon determined to move to Virginia and establish a full-fledged nursery in addition to keeping horses in training.

A number of buildings were already on site at Catoctin: the main house, the central block of which dates from the eighteenth century; a smaller dwelling, which is now the farm office; a number of frame barns readily remodeled as stables; and a particularly handsome and massive stone and frame barn, which has been altered to house racehorses out of training. A major addition to the earlier structures is a fourteen-stall broodmare barn, designed by Thomas W. S. Craven, partner in the Charlottesville architectural firm of Johnson, Craven, and Gibson. This, like the older buildings, is an inside barn, the aisle closed off with double sliding doors, but the loft extends over a shedrow, and the stalls open both ways. All interior woodwork, upstairs and down, is clear oak. The exterior is painted redwood, and the building is covered by a heavy terne roof. The gable ends are cut back to give continuity to the shedrow, so that the hay doors are set in pedimented dormers, and eyebrow vents break the roof on either side. The aisles are floored in asphalt, the stalls in clay. The stall module is fourteen feet square, but the two foaling stalls are fourteen by sixteen, and between them is a twelve-by-fourteen observation room. Inside, the building is light, airy, and appealing. The exterior architecture is a gentle play on farm-barn configuration, embellished by a silo lobby let into the shedrow. This high round chamber is lit by three doorways with ocular windows above them. The conical roof is capped by a cylindrical martin house supporting a whimsical weather vane; this utilitarian flourish, executed in brass and steel, takes the form of the proverbial stork bearing a swaddled foal, like the one at the National Stud at Newmarket, England.

Whether brought by the stork or fetched from the sales ring, the Catoctin horses have brought the Stud considerable renown. The astonishingly successful pair of 1973 foals, Honest Pleasure and Optimistic Gal, purchased by the Firestones at the 1974 Saratoga sales, foreshadowed the 1980 triumphs of Genuine Risk, also bought at Saratoga. Catoctin annually offers its entire yearling crop for public sale.

The Catoctin racing string is trained elsewhere, but a number of Thoroughbred establishments include, like Sagamore, breaking, training, and the care of lay-ups in their programs. Kentmere Farm, in Rancho Santa Fe, near San Diego, is one such multipurpose operation. Under the name Kama, the farm gained fame after the war as a Saddle Horse operation, developed by Mrs. Cummings Miller, daughter of Jerome Kern. It is now owned by Mrs. Patricia Beck of Dallas, and a training track and various new buildings have been added to meet changed requirements. The layout is a model of up-to-date management systems suited to the southern California climate. The largest barn, for horses in training, contains twenty-eight stalls. There is a two-stall stallion barn

"Bull-pen" exercise arena at Kentmere Farm, Rancho Santa Fe, California.

Pedestal sun shelter at Kentmere Farm.

160

with breeding shed, an eight-stall broodmare barn with stalls hooked into a master television console, and six isolation stalls. All buildings are wired for sound and are monitored by a watchman. The circular exercise pen, its slatted walls sloping out under an umbrella roof, is noteworthy for its monorail hot walker. The leadshanks hang from a high circular channel under the roof, so that the entire area is free of obstacles, thus disposing of the most cogent argument against the use of mechanical in-hand exercisers. These ceiling monorails are still uncommon and expensive, but are reported to be comparatively immune to breakdown. They are certainly eminently safe.

The paddocks at Kentmere employ the familiar southwestern sun shelters, but in this case some of the shelters lie across the farm roads, extending over the fences on either side; rain or shine, feed racks, watering troughs, and the trucks that supply them are protected.

Stallion quarters at Kentmere Farm.

River Edge Farm, Buellton, California, an innovative Thoroughbred facility, was built in 1976 to the joint design of the owner and the manager. The three long barns converge on a central block which includes the office, the breeding shed, and general work area. Every aspect of efficient stable management has been considered in the layout.

Turkey Hollow Farm, Southern Pines, North Carolina. The high, plain inside barns of this Thoroughbred farm are set in a pecan grove. The income-producing trees shade the pasture in summer, dropping the effective temperature by ten to fifteen degrees; in winter the sun is unimpeded by the high leafless branches. Note the "rubber" fencing. Cut from fanbelt fabric, and stretched by tractor, this resilient material makes up for the labor involved in its installation by the protection it offers to skylarking youngsters.

162

The Kentmere stables are conventional inside barns, exceptional only for unusually spacious stalls where the tailboards necessary for the Saddlebreds have been removed. They would be serviceable in any part of the country, and indeed such buildings appear from coast to coast and from Maine to Florida, but the Ocala training stable of Mr. and Mrs. Thomas Lavery could only be found in a benign climate, for it consists of a double row of twenty-four Port-A-Stalls, under a high pavilion roof that covers a railed aisle surrounding the shedrow. Originally set up as a temporary expedient until more permanent stabling could be erected, it has proved so well adapted to local conditions that no changes, save for the addition of nearby storage facilities and staff quarters, have been made or are contemplated.

A Thoroughbred stable at Live Oak Stud owned by Dr. and Mrs. John C. Weber. It was designed by Mrs. Peter Widener in the early 1960s. This is an attractive interpretation of a traditional southern stabling system.

The training stable of Mr. and Mrs. Thomas Lavery, Ocala. A trim blue and white color scheme adds to the visual appeal of the simple but satisfactory layout.

163

In chilly weather the stalls may be closed and still receive ample ventilation from above; most of the year they are fronted with mesh screens. The roof is well pitched, and is grounded, so that the rain and lightning that come with the violent local electrical storms flow harmlessly off down the hillside; it is formed of translucent plastic panels that admit plenty of light in spite of the low eave-line that keeps both rain and direct sunshine from the stalls. The building is, in effect, a standard Thoroughbred training barn reduced to the barest effective essentials.

Interior of the riding arena at the Foxcroft School, Middleburg, Virginia.

An open schooling arena is of a type increasingly popular in the South and Southwest. Angled girders and broad plain roof make a serendipitously strong and graceful architectural statement.

The growth of the prefabricated building industry has provided many attractive options to the stable owner. Huge indoor schools or open, roofed arenas, and inside or shedrow barns of any desired size, with pitched, gabled, or monitor roofs, may be combined with stall modules of varying dimensions to create practically instant facilities to order, in a bewildering choice of configurations. Properly erected according to manufacturer's specifications, these structures are comfortable, safe, and efficient. In general, of course, their cost is far below that of similar buildings constructed by traditional methods. They are to be found in large breeding and training establishments, commercial stables, show stables, and fairgrounds, and, of course, in innumerable backyards. It is also possible to buy plans for barns or stables; components for building to such designs may often be obtained from the same source. The advantages of these structures are obvious; unfortunately, they are often wholly uninspiring in appearance.

165

E. R. and G. P. Boyer's Big Sky Ranch, Scottsdale, showing the later prefabricated buildings. Note the pipe corrals.

The juxtaposition of old and new in an expanded establishment, however, can attain a special charm. At E. R. and G. P. Boyer's Big Sky Ranch, near Scottsdale, the original barn is slump brick, inside and out. The twelve stalls are little adobe rooms, open to the aisle over low walls, and the doors are stanchion gates. It is an earlier and humbler version of the barn at Lazy Acres, but modified in that half the stalls also open into outdoor paddocks; like the Haskell stable, it is cool, light, and well ventilated. Little Huey, a Quarter Horse stallion of classic conformation and confiding disposition, presides over the ranch from his corner chamber. As the operation has grown, additional facilities have been required. These include two training rings, two hot walkers, a row of open-air concrete washracks, a bull pen by Port-A-Stall, and five prefabricated inside barns by various manufacturers. The newest of these, by Ready-Stall, has proved signally satisfactory, largely due to the heavy intrinsic insulation in walls and roof; the manufacturer has drawn a lesson from the old masonry systems. The combination mangers and hayracks in the stalls, constructed of heavy-duty rod and sheet metal, are of the same pattern as those in the nearby pipe corrals.

166

To eyes accustomed to green pastures, wooden fences, and three-sided run-in shelters turned against the wind, the ubiquitous southwestern pipe corrals seem casual, even makeshift. In fact, they work very well. They constitute a sturdy and serviceable surrogate for hard-to-find timber, and have the additional advantage of offering no inducement to the cribber or the tail-rubber. Fitted with sun shelters, feed racks, and watering devices, they keep their inhabitants in obviously healthy content. Water is often provided in steel drums (from which a vertical segment has been sawn) lashed to the top rails. These have a large capacity, and the curve of the cylinder cuts down on wasteful splashing. The pipe corrals at the Boyer ranch are arranged in blocks, with a shelter over the four corners at the center. Each block is surrounded by a passage wide enough to take a service vehicle. They are policed by a portly goat, which combines in its person the roles of scavenger, counselor, and friend.

Pipe fencing at Stanford. See also color plate 12.

167

Cactus fencing at the Hugo Reid Adobe. See also page 26.

168

Pipe fencing may also be seen on dog-run stalls and behind suburban homes; in some instances the piping is reinforced or replaced by a flourishing cactus hedge, as at the Hugo Reid Adobe. This modification is likewise frightening to the unaccustomed observer, but in point of fact seems to work very well indeed, perhaps because most of the horses thus confined have a large proportion of Quarter Horse blood. The Quarter Horse is, by and large, an eminently sensible animal, with a strong sense of self-preservation; certain prejudiced parties hold that these simple barriers might prove disastrous to a "hare-brained" Thoroughbred, Arabian, or Saddle Horse.

This is a question the authors are not qualified to address; certainly pipe fencing plays little part in the arrangements at Kentmere, Lasma, River Edge, or Somerset. A degree of luxury, however, often forms part of the background for certain breeds. At the Cynthia Wood Stable in Santa Barbara, California, a thoroughly businesslike and efficient operation, American Saddle Horses are bred, trained, and fitted for show or sale in particularly gracious surroundings. The design of this stable complex is striking, its arrangements ingenious and practical.

One of the twin end pavilions at the Cynthia Wood Stable, Santa Barbara. See also color plate 28.

The twenty-five-stall main barn is in a nicely balanced Modern Mission style, with stark white walls, picked out in dark wood, and reddish tile roofs. The building is set off by a sweeping oval of creamy sand and brown post-and-rail fencing. Hexagonal bullpens close the training aisle at either end. With doors shut, they become individual exercise or longeing areas, and when the doors are open a horse in harness can swing a vehicle through them and around the pen without breaking stride. Hexagonal offices and utility rooms frame the cross-aisle entranceway; all hay, feed, and bedding are stored in a separate shed building to the rear, the roof of which is supported by two enclosed polygonal chambers. There are two grooms' apartments on the lower level and a guest apartment upstairs. The eleven-stall stallion barn, sand ring, and residential quarters are far enough away to preserve the balance of the central composition.

The American Saddle Horse, like its cousin the Tennessee Walker and the Standardbred, Morgan, Quarter Horse, and certain color breeds, had its origin

Copper Coin Farm, Simpsonville, Kentucky. The farm's American Saddlebred horses are schooled and sold from this barn. The nearside section, windows shaded with bright awnings, contains the stall ranges; the other is an indoor arena. Offices and utilities are contained in the "hyphen."

Mrs. J. A. Smith Hackney Pony stable, San Luis Rey Downs, California.

in this country. Thoroughbred, Morgan, and Arabian ancestors appear in the bloodlines of this elegant creature, which was, as the name implies, originally developed to provide dependable, comfortable, and handsome transportation for persons who spent most of their waking hours in the saddle. Today, these much refined animals are particularly popular in the show ring, where the emphasis on brilliance tends to disguise the degree of hardy athleticism implicit in the high, collected gaits displayed. In the Saddlebred, as in the Walker, the show gaits are exaggerated and often cultivated by means of a long foot and a heavy shoe; but the muscular resources that produce this snappy and balanced action, and the docile and attentive disposition that permits an animal looking fit to jump out of its skin to respond without resistance to rein or leg, make the Saddle Horse a useful performer in other fields. It must not be forgotten that the breed was developed for general utility; this versatility has not been lost.

Kentucky was the cradle of the American Saddle Horse. Washington Denmark, considered to be the original foundation sire, was Kentucky-bred, as was

171

his dam, sired by the pacer Bald Stockings (said to be of Narragansett descent). A second line descends from Mambrino Chief II, also Kentucky-bred, who traces back to Messenger, Thoroughbred progenitor of the Standardbred. The Bluegrass State is still the heart of the Saddle Horse industry, but many distinguished breeders and trainers are now to be found elsewhere.

Hackney ponies are often popular with Saddle Horse fanciers, for they share the inborn stage presence and elastic athleticism of their larger relatives. It is a long time since they were in demand to add to the consequence of a lady's town equipage, but they are still seen in the show ring, and of late years have

The Del Cody barn, Liberty Bell Park, Philadelphia.

172

performed with great credit in Combined Driving events. Their courage and vivacity, which is attended by an uncommon toughness, continues to attract admirers.

The Mrs. J. A. Smith stable at San Luis Rey Downs, not far from San Diego, was built in 1973 to house Mrs. Smith's Hackney Ponies. Mrs. Smith showed her homebreds across the United States and Canada for many years with great success. Advancing age somewhat curtailed her journeys, but she continued to drive Lady Alicia, a fiery little black mare, to a number of California triumphs until very lately. When this combination of champions left the arena at the Del Mar Horse Show for the last time, it was to a standing ovation, and they took with them their seventh consecutive Grand Championship.

Mr. Smith founded the San Luis Rey Downs community, which is built around a Thoroughbred training center and a golf resort. The cruciform pony stable, set above sloping pastures, includes an owner's apartment, originally intended as a pied-à-terre, but now almost a full-time residence. The pleasant Mission style of the building assures a cool and airy interior. The extra-wide aisles are tanbark, but sections of brickwork have been laid in the cross-aisles so that animals going in and out must traverse them. This decorative touch has solved a vexing problem. Although show animals perform and train on yielding sand or tanbark, they must of course learn to accept the sensation and sound of going on a hard road. An occasion when a novice spooked dangerously upon first encountering pavement persuaded Mr. Smith that the easiest way to accustom his animals to such a change of footing and to a reverberating surface was to make it an unremarkable part of their daily lives.

The stabling at the San Luis Rey Downs Training Center, although modern in construction and materials, differs little from other track and sales barns. Training tracks have rows of inside barns, but most track and sales stabling is conventionally of the shedrow type. Some contemporary stables of this nature are put together out of Y-shaped concrete modules set tip-to-tip in adjoining rows. These stalls are fireproof, and lend themselves readily to extension, but the traditional shedrow, its deep shady aisle supported on plain posts, is still the most familiar configuration. Wide-aisle inside barns are found at some Standardbred tracks where training goes on throughout the year; but the track stalls tend to conform to the familiar pattern. The Del Cody stable at Liberty Bell Park in Philadelphia, put up in the 1960s, is pleasantly typical; a cinder-block range, with the shed closed off at the ends. The aisle is wide enough to permit harnessing there. Cross-tie rings are in the stalls, which are between ten and twelve feet square, dirt-floored, and deeply bedded in straw.

Bedding, it should be parenthetically observed, is even more diversified than flooring materials, varying with locale, expense, and preference. Straw, bluegrass hay, shavings, sawdust, peanut hulls, tobacco stems, beet pulp, bagasse, and even dried leaves are all used, although great care must be taken that a horse dosed with vermifuge cannot nibble at tobacco stems, since the combination is highly toxic, and many horses are allergic to certain shavings (particularly cherry, and even more particularly black walnut, which can be fatal). The very latest thing in bedding is shredded newsprint. It is clean, fairly dust-free, and, where it is available, competitive in price. Its chief drawback is the sour odor which it develops after contact with urine, an unpleasant addition to familiar stable smells.

For centuries straw was the preferred, indeed the obvious, bedding, but modern combination harvesters destroy straw in the threshing process. Nevertheless, although good straw is expensive and not always easy to come by (a great deal is imported from Canada), it is a necessity at tracks and raceways. The contractors who remove stable muck for use as fertilizer cannot accept any alternative. At Liberty Bell, the manure goes to Pennsylvania mushroom growers to play its part in a delectable recycling process.

The owner-designed barn for a small Standardbred breeding and training farm, with training stalls, foaling stall, and generous storage space.

174

In the background of the illustration are the motel-style Liberty Bell grooms' quarters. This convenient residence, combining comfort with proximity to the job, is a great boon to staff. Swift access to the stable is vital in an emergency. It is also a great help in everyday management, so that it becomes less and less unusual to find a home and stable under one roof. In the Moore Hunt country around Southern Pines, and in the Aiken area, a sizable number of these dual dwellings, both private and commercial, have gone up in the last twenty years. Some are on one level, like the Smiths' arrangements at San Luis Rey Downs; others are two-storied, with horses below and humans above, as at the Cynthia Wood Stable.

Camelot, built in 1966 by the writer Raymond Wolff, is now Libby Evans's hunting box. The attractive black-and-white quadrangle stands at the end of a drive that runs between pine-shaded pastures. The house is on the left. Eight Dutch-door stalls form the other two sides, with a tack and feed room in the corner. The stall nearest the residence has been converted to a grooming stall; it is often unrestful to sleep next to an occupied stall. In the hunting season, when the stable is full, it is an exacting job to keep horses, tack, and barn in trim condition, but the work is facilitated by the convenience of the layout.

Camelot, Southern Pines, North Carolina.

Full Cry Farm, Southern Pines.

Full Cry Farm, a commercial stable outside of Southern Pines, is an example of the "up-and-down" school. This teaching and boarding stable was built in 1973 by Mr. and Mrs. Michael Russell, to their own design. The apartment is centered in the block, between twin lofts. The stable area contains sixteen big box stalls with centered sliding doors; the utility rooms are clustered in the middle below the apartment, giving additional ceiling height to the family quarters; there is a wash aisle with heat lamps above it. A great deal of thought and, of course, labor, has gone into this ample but compact facility. It is the sort of barn where everything is as close to hand as may be, and in just the place one would first look for it.

The hunting box at Mr. Joseph Bryan's Sandy Lane Farm, also in Southern Pines, is an especially fine example of the two-story scheme. A handsome penthouse apartment shares the second floor with a big loft, its curving laminated beams exposed, over a splendidly equipped inside stable. Because of the extra

176

Tennessee Walker show stable near Tuscaloosa, Alabama. The owners' country pied-à-terre is a comfortable apartment above the stable. The wide aisle of the barn is deeply bedded to protect the pasterns of its long-footed occupants.

The hayloft at Sandy Lane Farm, Southern Pines. See also color plate 29.

177

width of the building dictated by the dimensions of the overhead flat, the aisle is unusually broad and the twelve stalls generous. There is an attractive staff apartment on this lower level, a feed room and a tack room, convenient equipment storage, and a large center-drained washroom and dispensary, which also holds the washer-dryer for blankets, bandages, and saddlepads. In addition to overhead lights, this room is equipped with sealed underwater fixtures set low in the walls to provide safe and full illumination under the dampest circumstances.

Fireproofing is extensive. The building is brick and frame, after the old Virginia style, but the modern Gothic mow and the owner's flat are sealed off by concrete floors and fire doors from the stable and from each other. There is no wood in the stable area, except for the stall doors. From the road it might be taken for a commodious but otherwise unremarkable low-pitched barn, for even the chimney has been disguised as a ventilator; but in fact the building is a model of sophisticated design.

Rancho Santa Fe, in southern California, is less well known as a horse-oriented community than Aiken or Southern Pines, but this wealthy and conservative enclave is also ranch country and supports, in addition to citrus groves and such professional operations as Kentmere (pages 160–161), an active riding club and any number of domestic stables. Most of these are simple and conventional enough, but others have been painstakingly designed to complement the houses they support. The most striking among the latter was built for Mr. and Mrs. Roland Sahm by the California architect Fred M. Briggs to chime with the style of their contemporary residence of his design. Mr. and Mrs. Sahm have a deep interest in Indian crafts and artifacts, and the towering interior spaces of the house, enclosed in poured concrete walls and redwood panels, are designed to set off by their angular monumental quality the shapes and textures of the Sahm collections. The stable stands near the bottom of the winding entrance drive, and is a part of the view from the conversation area around the living-room fireplace. The severe projecting monitor and tall doors echo the tone of the main structure. Within, the building is relatively conventional, save for the soaring aisle and the glassed gable ends. The single row of stalls opens into a paddock as well as to the aisle. The side facing the residence is occupied by the feed room, tack room, utility area, and lounge; additional hay may be stored above the stalls. The architectural statement of the building has not been allowed to interfere with the pragmatic arrangement of the well-ventilated, light, and workmanlike interior.

The challenge of incorporating subsidiary buildings into an architectural composition has stimulated architects from Mansart to Frank Lloyd Wright, and each has approached the problem in his or her own way. Wright, commissioned to provide supporting kennels and stabling for a South Carolina plantation house of his design, supplied a typically idiosyncratic solution. The Wright buildings at Auldbrass, on the Combahee River, blend with the towering moss-

A hunting stable in Aiken. The adjoining carriage house has become the owner's residence.

The contemporary stable of Mr. and Mrs. Roland Sahm, Rancho Santa Fe.

"AULDBRASS" YEMASSEE

FRANK LLOYD WRIGHT

hung live oaks, dark cypress swamps, and low-lying fields of the surrounding country. All exterior walls are slanted, with their untreated cypress planking laid diagonally; in the house and gatehouse clerestory windows cut in an Indian-arrow motif run under the eaves; the arrow design is repeated in the supports of the roof overhang, and in some of the Wright-designed furniture and equipment. The whole composition is based on a series of hexagonal modules.

180

The shedrow stable describes two sides of an unfinished hexagon between six-sided structures at either end. The kennels and stable are part of an extended service complex facing the house; the buildings are pierced by a covered entranceway. The stalls themselves have centered gates in half-walls under a generous overhang, and alternate planks in the rear walls are hinged so that they may be left open for ventilation. Feed, tack, bedding, and equipment are stored in the module at the end of the range. Like most Frank Lloyd Wright buildings, these are designed to be looked at, not written about; and because of the great trees, they are difficult to photograph.

These stalls were never intended for all-season occupancy. They are not large, and their gates, which open inward, do not have a full swing because of the angle of the wall, so that one might have difficulty in leading a large or recalcitrant animal in or out. The hinged plank ventilators are ingenious, but produce a strong draft. The deep eaves are shady, however, and the range is convenient to pipe paddock and to service rooms. Admirers of Wright's genius are accustomed to accommodating themselves to the sculptural quality of his work, but horses, extreme reactionaries that they are, might feel more at home in quarters of a less adventurous design.

The considerable advantages of the closed polygon as a stable form are increasingly appreciated. Governor Sharpe's octagonal stable (page 43) has had many successors, some of very recent date. An interesting private stable in central Alabama combines an octagonal stall area with an attached hexagonal service chamber. The stalls were designed to open both to the center and to dog-run paddocks and are floored in heavy rubber openwork mats over wooden gratings, which allow free drainage to a crushed rock base. This flooring is not often seen, as the mats are more cumbersome to handle than had been anticipated, but it is an interesting experiment in improved sanitation. The exterior of the building is weathered gray siding, and it contrives to look austerely functional and rather inviting at the same time.

There is another Alabama octagon barn not fifty miles south of the one discussed above, although the owner-designer was unaware of its similar neighbor when she built it. It also contains six inside stalls. It was planned for easy maintenance by a working "horsewife." The center is open to the high lantern;

Stall end of the two-polygon Alabama stable.

A new octagonal stable in Alabama.

182

the open mow is over the stalls, and is reached via the hay door, or by means of a circular iron staircase. (There is also a rope, installed by the children of the family to facilitate swift descent, but this addition does not form an integral part of the design.) The floor is dirt, and the feed, utility, and tack areas flank the single arched door. The building is pleasing in appearance and extremely light and airy; the admirable plan constitutes a major laborsaving device.

For large installations, the inside barn is probably more convenient than the shedrow in terms of stable management, but owners have always been alive to the pleasure of seeing their horses over the half doors of outside stalls while approaching the barn, or even without leaving the house. There are few climates in which Dutch doors may not be left open at least during sunlit hours almost every day of the year. In winter whoever has charge of the barn may miss the shelter of a closed space, but even that difficulty can be obviated by the addition of a rear aisle. If there is no interior aisle, the shedrow will generally be blocked off at one or both ends by feed and utility rooms, which also

The meticulously groomed pastures and Umbaugh stable of Mr. and Mrs. Leo Hamel, Litchfield, Connecticut.

183

Mr. and Mrs. Paul Fout's domestic stable.

Mr. William Brainard's hunting stable, now owned by Mr. G. T. Ward.

Washrack under the shedrow at the Brainard stable.

give some shelter in severe weather. This was the pattern followed by Mr. William W. Brainard, Jr., and Mr. Paul Fout, both eminently knowledgeable Virginia horsemen, when they designed their residential stables; in the more rigorous climate of northwest Connecticut, Mr. Leo Hamel has included a rear aisle in his double Umbaugh building, one of the most attractive stables in the area.

One of the most interesting manifestations of the current popularity of equines for recreation is to be found in the proliferation of modest horse-oriented community enclaves from coast to coast. Indian Springs, near Birmingham, Alabama, is one noteworthy example. Deeds and restrictions, as in Aiken or Southern Pines, have been carefully drawn in favor of the horse and its owner. There is a barn in every yard, and most are very snug and attractive. Fences, all carefully maintained, are fitted with ingenious "creeps," so that people may pass through but horses may not. Indian Springs supplies many members to the nearby Cahaba Pony Club; animals and stables have an air of Pony Club "T.L.C." about them.

Gilcrease Hills, in Tulsa, Oklahoma, is a planned residential development in which much attention has been paid to recreational resources. There are a pool and tennis complex, paved bicycle paths threading the area (using underpasses to avoid road crossings), and an extensive network of adjacent riding and hiking trails. There are no private barns here. Neighborhood horses board in a simple stable laid out in twin facing shedrows. Feed and hay storage areas are normally situated, but there is no tack room; each stall has a lockable tack and utility closet opening out of it, a practical arrangement that undoubtedly contributes to a feeling of good fellowship among the owners.

Val-E-Vue, an unassuming housing project near Springfield, Illinois, is a "horsey" community of a different complexion. The developer, who raised Shetland ponies, hit on the idea of presenting a pony to the purchaser of every lot. The gesture was a huge success. There is now a sizable common pasture with a big hay shed, where all the animals spend the summer; they are sent to winter elsewhere. A few backyard barns have gone up, but, in general, the animals are community property and change proprietors as the age, size, and skills of the riders change. It is enchanting to see the children pile off the school bus on a bright fall day and rush to greet their Thelwellian friends.

A logical successor to these almost accidental horse-oriented enclaves is that contemporary phenomenon, the planned residential center for equestrians. It is not certain where the first of these was established; perhaps the concept occurred to several inventive developers at the same time. Every homeowner is a member of the association that owns the stable, indoor school, and adjacent facilities, employs the manager and grooms, and maintains the rings and trails. Residences in such specialized subdivisions tend to be rather elaborate; the stable facilities are generally commodious, well planned, and constructed from prefabricated elements as cheaply as is consonant with health and safety. The stark pragmatism of the stabling has no effect on any individual domestic elegancies,

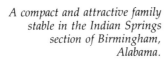

Gilcrease Hills community stable, Tulsa, Oklahoma, with a lockable tack closet built into every stall.

A compact and attractive family stable in the Indian Springs section of Birmingham, Alabama.

and the shared staff and consolidated operation relieve the owners from any barn work they cannot or do not choose to undertake. This revolutionary concept in mass housing to satisfy specific tastes, like many another brilliant notion, seems positively inevitable once someone has thought of it.

California is particularly well endowed with backyard horse areas. Zoning ordinances are liberal in regard to equine habitations. Some sections of suburban Los Angeles appear to have a horse in every yard; a simple fence and open shelter are, after all, the only requirements. In the northern part of the state, around Novato in Sonoma County, for example, there are likely to be two or more horses for every residence, for this is one of the loveliest trail-riding areas in the country. The shelters are of course more solid in character than in the warmer south and run the gamut from a prefabricated tool shed into which a Dutch door has been introduced to somewhat pretentious cupolaed barns, roofed and painted to match the house; but in general the stables of this rather prosperous horse-minded area reflect the growing trend toward pragmatic simplicity in equine housing, without much regard for traditional taboos.

In the hills above Novato, Mr. Woodward Melone has built a country retreat where he can indulge his love of horses and the out-of-doors. Mr. Melone,

Mr. Woodward Melone's two-stall bank barn, Novato, California.

a polo player, foxhunter, and horse lover, rides the surrounding back country daily, and is the founding member of a mounted fire-watch patrol. He has designed for his two trail horses a stable calculated to be at once trouble-free for himself and convenient for anyone else who might undertake the temporary care of the animals. It is, in effect, a bank barn on a full foundation, but the upper floor does not extend over the stall area, which is sheltered only by the roof. A pipe-rail partition separates the stalls and divides the pasture into sections, which may be thrown into one by opening a wide Texas gate. Hay and grain are stored on the upper level, which also contains a separate tack room. The horses are free to come and go at will, so that almost no stall maintenance is required, and water is supplied to the pasture. Hay and grain may be fed from above, and, from that vantage point, the horses observed rather closely for any injury or unsoundness. From the drive, the attractive building resembles a compact cottage, carefully sited to take advantage of the view. It has in fact been so planned that by extending the barn floor to the edge of the foundations, finishing the walls, and making a few electrical and plumbing connections, it could swiftly be converted to just such a purpose. The resulting residence might be delightful, but it would be a pity for such an expedient stable to be lost.

Venerable building systems and modern materials can combine with pleasing results in a pole barn. In Southern Pines, where a run-in shelter is all his tough little trotting ponies require, Mr. John Dean has put up the framework of a pole barn, roofed it in prefabricated panels, including two generous light screens on either side of the ridgepole, installed three-sided slatted stalls that open into a paddock, and achieved a harmonious and serviceable skeleton building. Because of the type of construction, he can enclose the barn, add further or more ambitious stalls, or extend the structure at will. Meanwhile, the ponies, the trailer, feed bins, and other essentials are adequately protected.

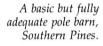

A basic but fully adequate pole barn, Southern Pines.

188

The simplest inside barn of all, on a hillside not far from San Francisco.

Mr. Dean's barn does not, however, represent the irreducible minimum in current stable design. That accolade must be bestowed on a California example, built on the brink of a freeway cut, somewhat below San Francisco. The gable roof covers an enclosed shed room on one side of an open alleyway and three stall pens on the other. The truncated stall partitions are laid up between the doubled metal poles that support the roof. There is no foundation, although the combination tack and feed room is floored. It too was built by the owner, and houses his two cutting horses, who appear well content with their simple quarters. In spite of modern components, both these barns have much in common with the stable buildings of medieval Europe or the pioneer shelter at Flower-dew Hundred.

IX *Children of Invention*

Who lets so fair a house fall to decay
Which husbandry in honor might uphold . . . ?
—WILLIAM SHAKESPEARE

The Sawmill River Parkway runs through what was, until not so long ago, the estate country of Westchester County, New York. There are still some superb homes secluded beyond the right-of-way, but the grandeur of the area is nowadays much subdued. Highways, subdivisions, and burgeoning villages encroach on the endless meadows where hounds of the Westchester Hunt used to show sport. Orchards are tangled and dying; one must go farther north now for cattle in the fields.

Up until the late sixties or early seventies a great half-timbered Norman stable complex, dating perhaps from the early twenties, was one of the fixed features of the view from the Parkway. It was laid out in a quadrangle open to the south, with one long side, supported on a high stone foundation built into a slope, facing the road. Horses used to graze in the pastures below it. Then they disappeared, and the surrounding acreage was laid down to hay; the post-and-rail fences were removed one by one. By the 1950s broken windowpanes had been replaced by plywood, which soon vanished in its turn, leaving dark blanks in the massive wall. Later, a mysterious fire severely damaged the fabric; shortly thereafter the building was razed. It is just possible to pick out the place where it stood by the granite remains of the foundation embankment.

The owners, their no doubt impressive residence, and the history of both, were alike anonymous to the traveler. The stable was sufficient unto itself as a landmark, an old friend in a familiar landscape. But when its size and location became a liability, and the last vestiges of its usefulness were destroyed by arson and decay, it was, very sensibly, pulled down. New generations negotiating the outmoded curves of the old Sawmill will not miss it. Hundreds of equivalent structures, confidently erected in anticipation of an untroubled future, have shared the same fate.

190

There are, of course, exceptions, some of which we have visited on this tour, and if a number of them lie under the threat of demolition, a few still echo to the chink of bits and the stamp of restless hoofs. Moreover, there are stables covering a wide spectrum of age, situation, and pretension, which have been for one reason or another preserved long after horses and horse-drawn vehicles ceased to occupy them. In country towns, many a backyard carriage house or stable, cherished for its storage capacity, acts as the repository for garden equipment, sleds and express wagons, lawn furniture, back issues of *Vogue,* the *National Geographic,* or *The Chronicle of the Horse,* dusty canning jars, rusted bicycles, musty steamer trunks, lame chairs and blind mirrors, limp decorations saved after a children's party, and all the rest of the flotsam that inevitably accrues to the best-regulated households.

Others, like the Black Horse Tavern at Flourtown or Governor Aiken's Gothic conceit (pages 49, 71), have been roughly pressed into service as car shelters by cutting an additional door, but are otherwise untouched. A conventional urban carriage house will hold a motorcar. The dimensions of the original doorway may present a challenge to the driver of a modern sedan, but the present tendency to regard smaller as better in automobiles is in a fair way to solving that problem. Downtown residential streets in older cities still show domestic

An anonymous backyard carriage house. Otherwise untenanted, it serves as auxiliary attic, cellar, and garden shed.

garages opening onto the pavement. Very few of these were constructed for cars. Even if they have been fitted with automated overhead doors, the masonry or metal wheel guards at the bottom of the doorframe betray their antecedents, and snow tires may yet be stored in a leftover tie stall.

Still other vintage stable buildings have simply been gutted to provide an open interior; these shells are presently workshops, machine sheds, or entrepôts for parks, museums, and even cemeteries (as at Sedgeley, Woodlands, or Vizcaya): thus, in a humble way, continuing to support the establishments for which they were erected.

None of these makeshift expedients, however appropriate or satisfactory, should properly come under the heading of adaptive use in the architectural sense, for this term implies a certain degree of creativity in the conception and execution of a recycling process. In truly successful adaptations, some part of the existing character of the buildings involved has been turned to advantage, so that they meet their new requirements with flair and perhaps in some instances better than they did their original purposes. Often the most ingenious and attractive stable conversions have been attended by the least sweeping alterations in structure or furnishing.

Certain new uses for equine housing have already become, in effect, traditional. Outside the cities, innumerable stables, carriage houses, and barns have been converted for residential use. In urban areas, the lion's share of domestic carriage-house–stables have been absorbed into the buildings to which they were attached (perhaps leaving a one-car space behind the carriage door) or have simply vanished into the foundations of some towering modern structure.

The Iron Gate. A popular Washington, D.C., restaurant, it was once the ample stable and carriage house for a downtown residence.

To survive as separate entities they must be put to uses sufficiently profitable to offset the soaring costs of urban land and air rights, which in turn means that enough of the original flavor must be preserved to convey an intriguing ambience.

This balance between expediency and nostalgia has been achieved to a nicety in certain city restaurants. A few rubbed and polished stall partitions left in place, or transferred to the original carriage chamber, suffice to break up a large plain room into pleasantly intimate subdivisions without impeding service. The kitchen is established in the smaller area, feed and tack rooms converted into offices or sanitary facilities, and the operation often made ready for business without any major structural revisions at all. The loft floor may be removed to improve ventilation and expose handsome old beams and trusses, or retained to provide additional dining space.

Carriage house and stable, Searles Castle, Great Barrington, Massachusetts, designed by Stanford White. Architecturally in keeping with its enormous turreted mansion, it is stylistically a far cry from the Italianate sophistication of Beacon Rock. Searles Castle, commissioned in 1882 by the widow of California railroad baron Mark Hopkins, has recently been opened as a house museum and setting for concerts, conferences, and exhibitions. The carriage house contributes solidly to meeting operating costs. There is a gift shop in the first-floor stable and a restaurant in the loft; an antiques mart fills the lower level, once devoted to spare vehicles, work horses, and farm equipment.

Idle employees used to be described as "living at rack and manger" at the expense of their employers, and one still occasionally hears the expression "putting on the feedbag" from persons who might be supposed never to have seen one of those handy equine lunch pails. Perhaps it is a sort of tribal memory of the contented rustling warmth in a stable full of horses, discussing their grain and hay in the dusk of a bleak winter evening, that underlies the appeal of dining in such an evocative atmosphere. Perhaps it is an association with the Dickensian groaning sideboards and steaming punch bowls of all those lovingly fictionalized hunt breakfasts. In any event, traces of stable fittings are carefully retained in more than a few remodeled eating places; other popular establishments, without a whiff of the stable in their pasts, contrive to suggest an apocryphal history of equine habitation.

This snug connotation may partially explain the attraction of the mews house or flat. When impressive town mansions lined block after block of city streets, their service buildings often occupied the interior of the block, and were reached by an alley around the corner from the big house. Some of these delightful anachronisms survive as residential cul-de-sacs, their smallish, narrow-

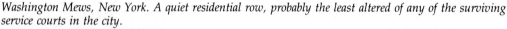

Washington Mews, New York. A quiet residential row, probably the least altered of any of the surviving service courts in the city.

194

Sniffen Court, New York City. Elegant maisonettes and a small theatre share this delightful backwater.

chested houses opening onto a paved or cobbled court. The only visible sign of their original use is likely to be a studio window in place of a hay door or a stone watering trough in the yard. Staircases are steep and rooms generally small, but a mews address carries a cachet (there are not many to be had) and offers a higher degree of privacy and relative quiet than may be looked for elsewhere in a crowded city.

Larger carriage houses and country stables may be more expansively treated for residential use. Some Victorian and Edwardian survivals are now sizable homes, their new interior arrangements varying according to the original plan, the needs of the family, and the predilections of the designer. All these buildings allow for a generous living room on the first floor, in addition to the usual offices; some are sufficiently spacious to include dining room, study, and downstairs bedroom as well. Stall partitions and wainscots are used to advantage to set off a kitchen area in an open plan or simply to modify a larger space.

Ingenious and comfortable country houses have been based on straight shedrow barns; part of the stall range can serve as bedrooms and bath, while larger chambers are gained by throwing two or three stalls together. The shedrow itself may be closed in, to provide an interior corridor, or left as a covered terrace, if the house is going to be occupied only in warm weather. Even a few grand establishments on the order of the Rolling Rock, Vernon Manor, or Pebble Hill stables have been turned into residences, often into a succession of flats

The Hennesseys, booksellers well known to collectors, operate from this nineteenth-century carriage house in Saratoga Springs, New York. The big corner box stall is the office. Its sturdy grill-capped partitions partially stem the tide of volumes which constantly threatens to engulf desk and files.

or maisonettes. Occasionally they form a single house surrounding an enclosed garden court, where part of the stabling may survive.

One spacious and beautifully furnished stable and carriage house has been turned into a party house for the use of an extended family. Here a short aisle of inside stalls has been retained in the rear of the building, "just in case" (color plate 21). Another, much earlier, building, its Victorian sensibilities ignored, has been metamorphosed into a poolhouse and cabana (page 202).

Of course, new roles for erstwhile stabling are by no means confined to restaurants or residences. In fact, the combination of imagination and pragmatism chronicled in a random list of remodelings shows that invention can make a virtue of necessity. The scope and ambition of these reworkings naturally depend upon the goals and financial positions of the owners, but a cheerful resourcefulness is the keynote of the genre. Among the facilities for which room has been found in the stable are shops, larger stores, dormitories, classrooms, studios, offices, galleries, showrooms, museums, theaters, at least one chapel, and even a gymnasium. These apparently superannuated structures have been turned to account in any number of ways. There is no reason to think that all options are exhausted.

When the farm surrounding these Amagansett, New York, barns was sold for development in the 1950s, the buildings were preserved by Stern Bros., of General Stores, as a department store. It is now the prosperous eastern outpost of the Brooklyn firm of Gertz & Co. Stalls and stanchions have been removed, but their supporting pillars, neatly boxed, serve as the focuses for various displays of merchandise; the hayloft contains the offices.

As social and economic changes have gradually relegated the vast, unwieldy, unheatable mansions of the past to dilapidated obsolescence, various organizations concerned with the preservation of architectural landmarks have undertaken the protection and maintenance of a wide selection of threatened behemoths. The first goal of such a project, of course, is to make the building as close to self-supporting as possible, and here the subsidiary stable can play a valuable part. Gutted, it may serve as a reception center, or, as previously noted, a storehouse, workshop, or machine shed. But if space allows, and it frequently does, more lucrative uses are possible. A gift shop in the stall area, a rentable assembly room in the carriage room, and perhaps a restaurant in the loft will generate considerable revenue from a relatively modest investment.

Below: San Diego architect Tom Hom transformed this large commercial livery stable into a farmers' market. The adaptation, which preserves an interesting building while filling a real need in the life of the city, received an "Orchids Award" from the San Diego chapter of the A.I.A.

The McGuire furniture showroom in San Francisco occupies the premises of the Zellerbach Paper Company stable, erected about 1860. Drays and wagons were kept on the ground floor, and the horses stabled above. Under the direction of Seattle architect Roland Terry, the wagon doors were replaced by mullioned display windows and the stable ramp by a staircase set at precisely the same angle. Mr. McGuire and his architect are to be congratulated on the practical preservation of one of the few surviving commercial buildings of this era in the city.

The Connecticut Free Library and Reading Room of Rowayton, Inc., designed by Edgar Moeller before the First World War as a sporting and farm stable for the James A. Farrell family. Eight hunters and polo ponies shared the first floor with stablemen and chauffeurs. Two stalls at ground level housed a succession of half-bred Percheron pairs, always called Nip and Tuck. The stall chamber, now the Community Center, was for many years a marine museum.

This drafting room in the Carlin & Pozzi office was the hayloft.

The stolid Victorian house in Schenectady, New York, which this building supported is boarded up and idle, but the subsidiary structure has been turned into excellent offices for the architectural firm of Feibes and Schmitt. The coach room is the drafting area, the carriage aisle a library, and the stable an airy conference chamber.

The Winder residence and the mews offices of Carlin & Pozzi, 3 Lincoln Street, New Haven. These adjoining buildings were cited for their successful demonstration of preservation through adaptive use.

200

The Kennedy Studios, on Beacon Hill, Boston, is delightfully housed in one of a pair of stables built in the 1820s by two brothers who could never agree. The Kennedy building was a livery stable, later became the headquarters of a riding and driving club, and has been so little altered that it could be returned to use as a stable with no difficulty whatsoever. The generous straight stalls, divided by iron posts and cap rails, serve as bins for prints, and the forge hearth in the rear still glows in supplement to the heating system. By leaving well enough alone, the firm has achieved a marvelous background for their framing business and a highly practical workplace.

201

Nathaniel D. Goodell completed this marvelously ornate Italian residence in Sacramento for Albert Gallatin, partner to Mark Hopkins, in 1878. For more than sixty years it was the California governor's mansion. The crested and pinnacled carriage house became an elegant cabana when a swimming pool was added to the mansion's amenities after the Second World War. The engraving, with its horse-drawn railway car, dates from the 1880s.

Hayracks, mangers, grain chutes, tie rings, and harness hooks easily establish a unified decor, and the elaborate grillwork of weighty stall partitions can support attractive displays of frivolous souvenirs. Indeed, existing stalls, or their remains, are often a treasure trove for the adapter. Box stalls may be sealed into rooms or left open as office cubicles or mini-boutiques. Straight stalls become booths or alcoves. If the building is too small for multiple use it can be made into an apartment for the staff or for independent tenants; where an uninterrupted space is required, the sturdy oak planking from the stalls can be used in new walls or interior finish.

Tracking down carriage houses, stables, and horse barns that have been successfully transmuted into quite unrelated creatures is a rewarding pastime. If, however, one's interest in the pursuit is sparked by a sentimental affection for the buildings as they originally stood, there is a bittersweet taste in the appreciation of even the cleverest and most elegant of transformations. Intransigently, one regrets their change of status even while admiring the ingenuity that saved them from demolition or decay. It is thoroughly satisfying, therefore, to discover a situation where the boot is on the other leg; where a structure created for quite another purpose has been rescued from oblivion by being turned into a stable.

202

The Next Move Theatre, Boston, once a stable for the Boston Mounted Police. It was a bank barn, and the stage is fitted into the lower stall area.

The Remington Theatre, Ilion, New York.

The carriage house and stable at Stan Hywit (1911–1915), the Frank A. Seiberling estate in Akron. Stan Hywit, a superb and extraordinarily late example of American Tudor Revival, is preserved and maintained as a house museum, largely staffed by volunteers. The assembly room in the coachhouse and the gift shop in the almost undisturbed stable provide considerable financial assistance to the project.

The carriage house of the "Unsinkable" Molly Brown's mansion, now operated by Historic Denver, Inc., has been converted into a roomy cottage apartment for the use of the director. Stone without and brick within, its exposed interior walls, heavy beams, and ocular or tripartite windows manage to combine Victorian and contemporary appeal.

Above and COLOR PLATE *25. Stanford White designed this impressive stable complex at Beacon Rock, Newport. The stall area is to the right; facing it are the onetime coachman's quarters and utility rooms, which have been turned into a charming residence. The carriage block, loft floor removed, is the studio of sculptor Felix de Weldon, creator of the Iwo Jima Monument. The arcaded loggia fronting the coach house has a ceiling embellished with striking Guastavino thimble vaults.*

COLOR PLATE 26. *A Massachusetts stable, designed by Eleanor Raymond in the 1930s to complement existing barns.*

COLOR PLATE 27. *Mission-style stable at Somerset Downing Arabian Stud in the Santa Ynez Valley, California. The adobe and tile block exterior of this 1968 building has been reproduced in a second barn, with interior fittings by Port-A-Stall.*

COLOR PLATE 28 *(overleaf). The Cynthia Wood Stable, Santa Barbara, California.*

COLOR PLATE 29. *Sandy Lane Farm, Southern Pines, North Carolina.*

COLOR PLATE 30. *A tobacco barn converted to stabling, at William Floyd's Fairway Farm, Lexington.*

COLOR PLATE 31. *A Hansel and Gretel fantasy in stone, designed by Dudley Newton in the late 1800s, and formerly the stable for Belair, Newport. Now in the process of painstaking restoration and finishing by the present owner, it has become a charming two-family house.*

COLOR PLATE 32. *A Gilded Age nabob's stable and carriage house, as converted by J. Sadler Designs.*

Above: The nabob's stable and carriage house, before conversion. Below: the living room fills the upper stall range and, save for the insertion of a fireplace, the removal of a few stall partitions, and the refinishing of the woodwork, is virtually unchanged.

Above: A Connecticut stable, converted by the owners into a summer home amid the building restrictions of the Second World War. The sheltered inner terrace under the shedrow serves as the hallway.

Opposite
Above: "Horse Haven." (Left) Thomas Hastings fitted this modest house and stable into a fifty-by-one-hundred-foot lot in Aiken, thus not only winning a wager from a skeptical friend, but ensuring that his wife could observe the horses from the house, as she could at La Bagatelle (see page 122). (Right) The little stable building is now a guesthouse for the present owner, Dr. Robert Lipe, and is lined with paneling which he obtained from a local mansion under demolition. Photographed from the terrace, once the stableyard.

Below: This little house in Beverly Hills is another product of the post-war housing shortage; it was Rudolph Valentino's stable.

Our country landscapes are well-supplied with the remains of rural school-houses, shuttered and cold or tumbling to ruin, their bells silent, perhaps, for nearly a century. Unless preserved for their historic value or moved to new sites, they are generally the object of the attention of casual vandals only, but at least one four-square frame academy, repaired and suitably remodeled, is now a comfortable inside stable that includes a family recreation room in the plan, as well as all appropriate storage spaces; chalk and McGuffey's Readers are forgotten. A modern carriage museum has been built in such close imitation of a stable that it could readily become one; the scheme is not only appropriate but extremely workable.

Right: The stable at Lyndhurst, Tarrytown, New York, built to support the A. J. Davis Gothic mansion. Lyndhurst is now a property of the National Trust for Historic Preservation, which has placed a gift shop in the carriage house but left the stall area undisturbed; indeed, it has lately housed a smart Morgan belonging to one of the curators.

Below: Beaver College, Glenside, Pennsylvania, is situated on the William Welch Harrison estate. Modeled after Alnwick Castle, this "superb pile" was the first of Horace Trumbauer's large residential commissions, and led to his subsequent extensive employment by the Widener family. The stable building is a formal enclosed courtyard, pierced by sally ports, with a sizable carriage house at one end. The stall ranges to the left of the arch have been turned into craft studios, and above the classrooms in the carriage house is the nonsectarian chapel.

The 1926 stables at the Kellogg Arabian Ranch, now part of the California Polytechnic University campus. Although it was the original Kellogg stable and served as the focal point of the ranch for years, certain practical considerations were sacrificed for architectural effect. The angle of the roof fails to cut off the glare of the courtyard from the stalls, and the monorail manure carrier, designed to save steps, was generally out of order. The building is assigned to the use of the Student Union, and a number of interesting and venturesome plans have been made for its future use.

The Carriage Museum at Stony Brook, New York, built in echo of a carriage house and stable as a sensible and convenient setting for its collections.

There are other examples to be found, but the most marvelously unconventional adaptation is surely the auxiliary stall range built into the back of a set of disused concrete bleachers, one stall to each supporting arch. The unassuming occupants, looking out over rough half doors, are superbly framed — and shaded — by arcaded masonry. There is a delicious economy about this expedient, and a strong dash of poetic justice.

Rear view of the outmoded concrete bleachers next to the polo stables at Golden Gate Park, San Francisco. The stables themselves now house an active boarding and livery establishment, which, like many a similar operation, has found itself short of conventional stall space.

This rough (but ready) annex marks the formal conclusion of our stable tour, but by no means closes the subject. We intend to add to our "stable collection" whenever opportunity serves, and heartily recommend to others to do the same.

The vast area of our country is — it is not too much to say — positively fraught with stable buildings. Large or small, elegant or blockish, historic or contemporary, built for the ages or prefabricated for instant relocation, abandoned, adapted, or in active service, they are informed by a common charm. In the manner of all enchantments, this charm is not wholly logical. After all, in spite of wide variations in luxury and layout, a stable is only a stable. Like chicken houses, kennels, and milking parlors, stables are essentially created to domicile domestic animals safely and in a fashion convenient to their masters. The poultry farmer, dog breeder, cattleman, and horseman must always be interested in learning from the ingenuity of their peers, but a stable seems to have a special attraction quite apart from any workaday lessons to be learned from it.

A good stable, one in which horses are (or have been) healthy and well cared for, has a reassuring architectural personality that combines dependability and warmth. It is odd that a free-ranging, grazing animal, which in its natural state has no lair or den, and rarely seeks any enclosed shelter, should stamp

210

such a feeling of homely comfort on what is, in fact, "an habitation enforced," but that feeling permeates every well-ordered horse barn, and tends to persist while the building stands. This is the lagniappe that falls to those who, chiefly interested in the architecture or technology of the building, constantly rediscover the pleasures of visiting a stable.

Gone Away!

Acknowledgments

It is a real pleasure to write a thank-you note to the people who, in many and various ways, have contributed to the making of this book. The difficulty lies in setting forth their names in appropriate fashion. Logically, they should be categorized, beginning of course with the owners and their staffs who gave us free access to their premises and graciously permitted us to use their stable buildings; continuing with the professionals—librarians, curators, custodians—who have patiently and cheerfully put their resources at our disposal; going on to those who generously supplied illustrations, and concluding with the roster of true friends, old and new, who gave unstintingly of their time, local knowledge, eclectic expertise, and warm hospitality. But in this case the logical system breaks down, because so many names fall into two, three, or all of the divisions.

In alphabetical order, therefore, follows a list of all those who helped. No book has ever been created in a vacuum. Everyone here has had some hand in the evolution of this one, and we are grateful.

Mr. and Mrs. Steven Albernaz, the late Robert J. Alcorn, Mr. and Mrs. Donald J. Anderson, Neil Ayer, Edward Ayres, Mr. and Mrs. H. Parrott Bacot, Mr. and Mrs. John Bakhaus, Dr. David J. Balch, Arthur J. Ballard, Mrs. John T. Barber, James Barnes, Sonya Bay, Mr. and Mrs. Charles N. Bayless, Betsy Beach, His Grace the Duke of Beaufort, Patricia Beck, William Beiswanger, Sir Alfred and Lady Beit, David Bennett, Mr. and Mrs. Walter V. Bennett, Jr., Mr. and Mrs. Gilbert B. Benson, the late Thomas Berry and Mrs. Berry, Mr. and Mrs. James H. Blackwell, Robin Bledsoe, Judith Bloomgarden, Gray Boone, John Bowles, Mr. and Mrs. William Brainard, J. S. Brandt, Marguerite Brooks, Bennie Brown, Jr., Robert Breugmann, Joseph M. Bryan, Jr., Tom Budney, Maggie Burkley, John C. Burns, William Butler.

Margaret Cameron, Mr. and Mrs. George S. Campion, Mr. and Mrs. Hill Carter, Mrs. James J. Casey, John Castellani, William A. B. Cecil, Edward A. Chappell, Mr. and Mrs. Charles A. Chapin, Helen Chillman, Mrs. George P. K. Ching, Louise Christie, Cys Clark, Melissa Clemence, Mr. and Mrs. Hans Coester, Mrs. Thomas Cohn, Herbert Collins, Mr. and Mrs. Richard Sloane Colt, Georgianna Contiguglia, Lisa Cox, Dr. and Mrs. William Craddock, John Craven, Thomas W. Craven, Charles J. Cronan, Jr., Michael F. Crowe, Lady Cusack-Smith, Jay Dalgliesh, Jeffrey Darbee, Mr. and Mrs. John K. Dean, Gilbert J. Denman, Jr., Mr. and Mrs. Magruder Dent, Mr. and Mrs. Felix de Weldon, Deb Dows, Russell Drake, Charles H. P. Duell, Dale Duffy, Mary Dunnigan, Mr. and Mrs. Lee Eberle, Mrs. S. Henry Edmonds, John Evangelisti, Libby Evans, Stanley Falconer, Mr. and Mrs. Paul Fout, H. H. Ferguson, Mr. and Mrs. Bertram Firestone, Mr. and Mrs. John Fitzhugh, William D. Floyd, William Barrow Floyd, Ronald W. French.

Mr. and Mrs. G. William Gahagan, John Gaines, John Garcia, Donald Garretson, Mr. and Mrs. H. Williamson Ghriskey, Jr., the Hon. Rosemary Gill, Mrs. Roswell L. Gilpatric, Bernard Gnapp, James Goode, Dr. A. S. Gordon, Mr. and Mrs. Thurston Greene, Wendy Grieder, Milton Grigg, Mr. and Mrs. William Groff, Shelley Groom, the Hon. Desmond Guinness, Dr. Melvyn L. Haas, Karl Haglund, Nancy Haight, Mr. and Mrs. Sherman Post Haight, Jr., Mr. and Mrs. Leo Hamel, Edith Harrison-Conyers, Iola S. Haverstick, William F. Heins III, Thomas A. Heinz, Charles Hodges, Mr. and Mrs. Nicholas Holmes, Robert Hunker, Conover Hunt-Jones, Mr. and Mrs. W. Bradfield Hutchings, Mrs. Lawrence Illoway, Mary Ison, Mr. and Mrs. Walter M. Jeffords, Gregory Johnson, Mr. and Mrs. Joseph F. Johnston, Morgan Jones, the late Olwen M. Jones, Warner Jones, Martha Koon.

Dr. Eugene E. La Croix, Joe LaGuardia, Mr. and Mrs. Alvin Landau, Mills Lane, Mr. and Mrs. George E. Lane-Fox, Walter Langsam, Norman Larson, Mr. and Mrs. Thomas Lavery, Mrs. Thomas Law, Mrs. Harold LeBlond, Elizabeth Lewis, James E. Lewis, Mr. and Mrs. Leonard T. Lewis, Dr. Robert O. Lipe, Mr. and Mrs. R. K. Longchamps, Robert F. Looney, Mrs. Stanton D. Loring, Calder Loth, Mr. and Mrs. Al Louer, Dorothy Lyons, Carol McIntyre, Alexander Mackay-Smith, Robert MacKay, Mr. and Mrs. Edward Manigault, Esmond Martin, Mr. and Mrs. Benjamin Matteson, Edward McCloskey, Jane McIlvaine McClary, John C. McGuire, Mr. and Mrs. Thomas McLaughlin, Joan McLoughlin, Woodward Melone, Paul Mellon, Sir John Miller, John F. Miller, Herbert Mitchell, Willard B. Moore, Dee Muma, the Hon. and Mrs. David Nall-Cain, Dr. and Mrs. E. D. Vere Nicoll, Richard Nylander, Page Oberlin, Mrs. John E. O'Brien, Frances Campbell Orf, Mr. and Mrs. Henry O'Shaughnessey.

Richard F. Pappalardo, Mrs. David Parsons, Amelia Peabody, Mr. and Mrs. Norman Pease, James H. Peden, Ford Petross, David Pettigrew, Betsy Pitha, Dr. Joachim M. Plotzek, Walter Plunkett, Mr. Parker Poe and the late Mrs. Poe, O. Powell, Mrs. Igor Presnikoff, Mr. and Mrs. James W. Proctor, Mr. and Mrs. James Pyle, Mrs. Robert Rainey, Monica Randall, Maj.-Gen. and Mrs. Dilman Rash, Mr. and Mrs. Roger C. Ravel, Mr. and Mrs. James C. Rea, Mrs. John A. Reidy, Gabor Renner, Mr. and Mrs. Howard Richmond, Mr. and Mrs. James K. Robinson, Jr., Mr. and Mrs. John S. Rodes, Lee Romney, Guy Roop, Lee Russell, Mr. and Mrs. Michael Russell, Mr. and Mrs. John Russo, John C. Saah, Mr. and Mrs. John H. Sadler, Jason Sadler, Garrett Sadler, Sue A. Sadler, Mr. and Mrs. Roland Sahm, Mr. and Mrs. Charles Scarlett, Robin Scully, William Seale, Louise Serpa, Mr. and Mrs. Wallace Shanks, Mr. and Mrs. M. L. L. Short, Charles Simmons, Mr. and Mrs. J. A. Smith, Mr. and Mrs. Lawrence W. Smith, Mrs. L. Durbin Smith, Lucy Graddy Smith, Sandra Snider, Mr. and Mrs. William C. Steen, Pierre Stevens, Chauncey Stillman, Mr. and Mrs. Samuel Sutphen, Edward Swain III, Mr. and Mrs. Roy Swayze, Mr. and Mrs. George Sweetman, Countess Szapari.

Mrs. Cabell M. Tabb, William Taggart, Col. and Mrs. H. Gwynne Tayloe, Paul Taylor, Thomas Taylor, Warren Taylor, Carole Teller, David Terrell, Wayne Thomas, Mr. and Mrs. Richard M. Thune, Mrs. C. B. Thuss, Betsy B. Tierce, the Tinney family, Col. and Mrs. Cloyce Tippett, Judith A. Trotter, Douglass Shand Tucci, Margaret Tuft, the late Mrs. H. C. Turner, Jr., Kenneth I. Urquahart, Daniel S. Van Clief, Alfred Gwynne Vanderbilt, Gypsy Vanderveer, Wendy Wagner, Dr. Jacqueline Claus Walker, G. T. Ward, Lowry Watkins, Christopher Weeks, Mrs. W. H. Wemyss, Peg Whitehurst, Thomas Wilboldt, Reid Williamson, Dr. Calvin Wise, Cynthia Wood, Mr. and Mrs. Lloyd Wood, Mr. and Mrs. Neal S. Wood, Marjorie Wynne, Tom Young.

212

Index

214

Photograph Credits

The illustrations are reproduced by the courtesy of the following owners, photographers, and artists, or from the following sources. All photographs not listed here were taken by Julius Trousdale Sadler, Jr.

The color plates are found on pages i–iv and following pages 12, 76, 140, and 204.

The American Institute of Architects Foundation/Architectural Archives, Washington, D.C. (photo William Edmund Barrett): 92; Anheuser-Busch, Inc.: color plate 15 (both); *The Magazine ANTIQUES* (drawing Roy Frangiamore): 35; *Architectural Review*, Feb. 1902: 101, 128; John R. Barber: 77; Barberton (Ohio) Historical Society: 118, 119; Charles Bayless: 38 right, 49, 50 (both supported by a grant from the National Endowment for the Arts in Washington, D.C., a Federal agency); Beinecke Rare Book and Manuscript Library, Yale University: 13, 14; Town of Bowie (Md.) Museum: 113; Trustees of the British Museum (copyright British Museum): 17, 52 bottom; Robert Bruegmann: 95 top; *Bulletin of the American Meteorological Society*, Sept. 1960: 55; Eugene R. Burke, Cincinnati: 107; California State Library: 28; James Andrew Carr: 205 bottom; Narcissa Chamberlain (photo Samuel Chamberlain): 52 top; Jeannette Ching: 4; Cys Clark: 4, 211; The Condé Nast Publications, Inc.: 157 (copyright © 1939, 1967); 206 bottom (from *House & Garden*; copyright © 1944, 1972); 207 (from *House & Garden*; copyright © 1947, 1975; photo André Kertész; *Country Life* (U.S.): 2, 143; Hugh Doran, Dublin: 22 top; Olin Dows: 146 top; The Dunlap Society, Essex, N.Y.: 61; *Equus*: 102; J. W. Fiske Iron Works, *Stable Fixtures: Illustrated Catalogue* (1924): 144; Flowerdew Hundred Foundation: 24 top left; Foxcroft School (photo Winants Bros.): 164; Free Library Company of Philadelphia: 67, 71 left; Historic Charleston Foundation: 34; Historic Denver, Inc.: 204 bottom; The Historical Society of Pennsylvania: 69 bottom, 78 bottom left; Irish Georgian Society: 21, 22 bottom; Richard V. N. Gambrill and James C. Mackenzie, *Sporting Stables and Kennels* (1935): 142; Irish Tourist Board: 20; Robert Kahler (copyright © *The Stanford Magazine*): color plate 12; Thomas Jefferson Memorial Foundation, Monticello (photo Edwin S. Roseberry): 58; Liberty Bell Park, Philadelphia: 172; Library of Congress: 30 top, 38, 39 left, 48 (bottom: photo C. O. Green, 1940), 53 bottom (HABS drawing), 60, 66, 72, 83 bottom (HABS drawing), 85, 202; Los Angeles State and County Arboretum, Arcadia, Calif.: 87 right; The McGuire Company, San Francisco: 199 top; Massachusetts Historical Society, Boston: 57; Mount Vernon Ladies' Association, Mount Vernon, Va.: 45 bottom, 46 bottom; Musée de Versailles (photo Réunion des Musées Nationaux): 16; Museen der Stadt Köln (photo Rheinisches Bildarchiv, Kölnisches Stadtmuseum): 12; Museum of the City of New York: 79 bottom, 104, 105; National Archives: 27, 74, 75, 76 top; National Register of Historic Places: 203 bottom; National Sporting Library, Middleburg, Va.: 59 bottom (from *American Turf Register*, April 1840), 131; New Haven Colony Historical Society: 69 top; New York State Fairgrounds: 81; The Next Move Theatre, Boston: 203 top; Newport Historical Society: 99; Newport Preservation Society: 100; North Carolina State Department of Cultural Resources, Division of Archives and History, Raleigh: 40, 41 top; North Shore Preservation Society, Oyster Bay, N.Y.: color plate 23 (top); Ohio Historic Preservation Office: 117; Plimoth Plantation, Plymouth, Mass.: 24 bottom; Paul J. Pozzi: 200 top left; James C. Rea: 64 left; John C. Saah: 192; J. Sadler Collection: 7, 8; St. Joseph (Mo.) Museum: 78 top; San Antonio Conservation Society: 96 bottom; C. Scarlett: 43; Louise L. Serpa: 134; Silver Ranch, Inc., Jaffrey, N.H.: 89 bottom; Nancy Sirkis: 93; Society for the Preservation of New England Antiquities: 32, 33, 65; Stan Hywit Foundation: 204 top; Carole Teller: 63 left, 194, 195; Lindsey Thune: 93 top; Tryon Palace Restoration: 41 bottom; Douglass Shand Tucci: 70; Valley Forge National Historical Park: 24 top right; Mariana Griswold Van Rensselaer, *Henry Hobson Richardson and His Works* (1888): 95 bottom; The Virginia Historical Society: 36; Vizcaya Museum and Gardens (photo Doris Littlefield): 120; Elizabeth B. Wood: 64 right; The Frank Lloyd Wright Memorial Foundation: 180; Yale University Art Gallery. Whitney Collections of Sporting Art, given in memory of Harry Payne Whitney (B.A. 1894) and Payne Whitney (B.A. 1898) by Francis P. Garvan (B.A. 1897) June 2, 1932 (photo National Museum of Racing, Inc., Saratoga Springs, N.Y.): 59 top.

Endpapers: Edward Mayhew, *The Illustrated Horse Management* (1868).

Library of Congress Cataloging in Publication Data

Sadler, Julius Trousdale, Jr.
    American stables, an architectural tour.

    Includes index.
    1. Stables—United States.   I. Sadler,
Jacquelin D. J.   II. Title.
NA8340.S22   728'.9   81-3985
ISBN 0-8212-1105-6   AACR2